Data-Centric Systems and Applications

For further volumes:
http://www.springer.com/series/5258

Peter Christen

Data Matching

Concepts and Techniques
for Record Linkage, Entity Resolution,
and Duplicate Detection

 Springer

Peter Christen
Research School of Computer Science
The Australian National University
Canberra, ACT
Australia

ISBN 978-3-642-43001-5   ISBN 978-3-642-31164-2 (eBook)
DOI 10.1007/978-3-642-31164-2
Springer Heidelberg New York Dordrecht London

ACM Computing Classification (1998): H.2, H.3, I.2, I.5

© Springer-Verlag Berlin Heidelberg 2012
Softcover reprint of the hardcover 1st edition 2012

Printed on acid-free paper

Springer is part of Springer Science+Business Media (www.springer.com)

To Gail

Foreword

Early record linkage was often in the health area where individuals wanted to link patient medical records for certain epidemiological research. We can imagine the difficulty of comparing quasi-identifying information such as name, date of birth, and other information from a single record against a large stack of paper records. To facilitate the matching, someone might transfer the quasi-identifying information from a set of records to a large typed list on paper and then, much more rapidly, go through the large list. Locating matching pairs increases in difficulty because individual records might have typographical error ('Jones' versus 'Janes', 'March 17, 1922' versus 'March 27, 1922' because handwriting was difficult to read). Additional errors might occur during transcription to the typewritten list.

Howard Newcombe, a geneticist, introduced the idea of odds ratios into a formal mathematics of record linkage. The idea was that less frequent names such as 'Zbigniew' and 'Zabrinsky' (in English speaking countries) had more distinguishing power than more common names such as 'John' and 'Smith'. Among a pair of records that were truly matches, it was more typical to agree on several quasi-identifying fields such as first name, day of birth, month of birth, and year of birth than among a pair of records that had randomly been brought together from two files.

Newcombe's ideas were formulated in two seminal papers (Science, 1959; Communications of the Association of Computing Machinery, 1962). These papers contained a number of practical examples on combining the scores (odds ratios) from comparisons of individual fields in a pair to a total score (or total matching weight) associated with a pair. The combining of the logarithms of the scores via simple addition is under conditional independence (or naïve Bayes in machine learning). Pairs above a certain higher cutoff score were designated as links (or matches); pairs below a certain lower score were designated as a non-link (or non-match); and pairs between the upper and lower cutoff scores were known as potential links (potential matches) and held for clerical review. During the clerical review, the clerk might correct a name or date of birth by consulting an alternative source (list) or the original form (that might have had typographical error introduced during data capture to the computer files).

Obtaining the odds–ratios for suitably high quality matching would have been very difficult in most situations because training data were never available. Newcombe had the crucial insight that it was possible to compute the desired probabilities from large national files such as health or death indexes (or even censuses). He obtained the probabilities associated with the linked pairs by summing the valuespecific frequencies for individual first names, last names, etc., from the large file. He used the frequencies from the cross-product of the files along with an adjustment for those frequencies associated with linked pairs to get the appropriate frequencies for non-linked pairs. Newcombe's methods were robust with new pairs of files because the 'absolute' frequencies from the large national files worked well.

Fellegi and Sunter (Journal of the American Statistical Association, 1969) provided a formal mathematical model where they proved the optimality of Newcombe's rules under fixed upper bounds on the false link (match) rates and the false non-link (non-match) rates. The methods were later rediscovered by Cooper and Maron (Journal of the Association of Computing Machinery, 1977) without proofs of optimality. Fellegi and Sunter extended the model with ideas of unsupervised learning and extensions of value-specific frequency concepts with crude (but effective) ideas for typographical error rates.

For more than a decade, most of the methodological research has been in the computer science literature. Active areas are concerned with significantly improving linking speed with parallel computing and sophisticated retrieval algorithms, improving matching accuracy with better machine learning models or third-party auxiliary files, estimating error rates (often without training data), and adjusting statistical analyses in merged files to account for matching error.

Many applications are still in the epidemiological or health informatics literature with most individuals using government health agency shareware based on the Fellegi–Sunter model. Although individuals have introduced alternative classification methods based on Support Vector Machines, decision trees and other methods from machine learning, no method has consistently outperformed methods based on the Fellegi-Sunter model, particularly with large day-to-day applications with tens of millions of records.

Within this framework of historical ideas and needed future work, Peter Christen's monograph serves as an excellent compendium of the best existing work by computer scientists and others. Individuals can use the monograph as a basic reference to which they can gain insight into the most pertinent record linkage ideas. Interested researchers can use the methods and observations as building blocks in their own work. What I found very appealing was the high quality of the overall organization of the text, the clarity of the writing, and the extensive bibliography of pertinent papers. The numerous examples are quite helpful because they give real insight into a specific set of methods. The examples, in particular, prevent the researcher from going down some research directions that would often turn out to be dead ends.

Suitland, USA
William E. Winkler
U. S. Census Bureau

Preface

Objectives

Data matching is the task of identifying, matching, and merging records that correspond to the same entities from several databases. The entities under consideration most commonly refer to people, such as patients, customers, tax payers, or travellers, but they can also refer to publications or citations, consumer products, or businesses. A special situation arises when one is interested in finding records that refer to the same entity within a single database, a task commonly known as *duplicate detection*. Over the past decade, various application domains and research fields have developed their own solutions to the problem of data matching, and as a result this task is now known by many different names. Besides data matching, the names most prominently used are *record* or *data linkage, entity resolution, object identification*, or *field matching*.

A major challenge in data matching is the lack of common entity identifiers in the databases to be matched. As a result of this, the matching needs to be conducted using attributes that contain partially identifying information, such as names, addresses, or dates of birth. However, such identifying information is often of low quality. Personal details especially suffer from frequently occurring typographical variations and errors, such information can change over time, or it is only partially available in the databases to be matched.

There is an increasing number of application domains where data matching is being required, starting from its traditional use in the health sector and national censuses (two domains that have applied data matching for several decades), national security (where data matching has become of high interest since the early 2000s), to the deduplication of business mailing lists, and the use of data matching more recently in domains such as online digital libraries and e-Commerce.

In the past decade, significant advances have been achieved in many aspects of the data matching process, but especially on how to improve the accuracy of data matching, and how to scale data matching to very large databases that contain many millions of records. This work has been conducted by researchers in various fields, including applied statistics, health informatics, data mining, machine learning, artificial intelligence, information systems, information retrieval, knowledge engineering, the database and data warehousing communities, and researchers working in the field of digital libraries. As a result, a variety of data matching and deduplication techniques is now available. Many of these techniques are aimed at specific types of data and applications. The majority of techniques has only been evaluated on a small number of (test) data sets, and so far no comprehensive large-scale surveys have been published that evaluate the various data matching and deduplication techniques that have been developed in different research fields.

The diverse and fragmented publication of work conducted in the area of data matching makes it difficult for researchers to stay at the forefront of developments and advances on this topic. This is especially the case for graduate and research students entering this area of research. There are no dedicated conferences or journals where research in data matching is being published. Rather, research in this area is disseminated in data mining, databases, knowledge engineering, and other fields as listed above. For practitioners, who aim to learn about the current state-of-theart data matching concepts and techniques, it is difficult to identify work that is of relevance to them.

While there is a large number of research publications on data matching available in journals as well as conference and workshop proceedings, thus far only a few books have been published on this topic. Newcombe [199] in 1988 covered data matching from a statistical perspective, and how it can be applied in domains such as health, statistics, administration, and businesses. Published at around the same time, the edited book by Baldwin et al. [16] concentrated on the use of data matching in the medical domain. More recently, Herzog et al. [143] discussed data matching as being one crucial technique required for improving data quality (with data editing being the second technique). A similar approach was taken by Batini and Scannapieco [19], who covered data matching in one chapter of their recent book on data quality. While Herzog et al. approach the topic from a statistical perspective, Batini and Scannapieco discuss it from a database point of view. Published in 2011, the book by Talburt [249] discusses data matching and information quality, and presents both commercial as well as open source matching systems. Similarly, Chan et al. [51] present declarative and semantic data matching approaches in several chapters in their recent book on data engineering.

None of these books however cover data matching in both the depth and breath this topic deserves. They either present only a few existing techniques in detail, provide a broad but brief overview of a range of techniques, or they discuss only certain aspects of the data matching process. The objectives of the present book are to cover the current state of data matching research by presenting both concepts and techniques as developed in various research fields, to describe all aspects of the data matching process, and to cover topics (such as privacy issues related to data matching) that have not been discussed in other books on data matching.

Organization

This book consists of 10 chapters. Chapter 1 provides an introduction to data matching (including how data matching fits into the broader topics of data integration and link analysis), a short history of data matching, as well as a series of example applications that highlight the importance and diversity of data matching. Chapter 2 then gives an overview of the data matching process and introduces the major steps of this process. A small example is used to illustrate the different aspects and challenges involved in each of these steps.

The core of the book is made of Chaps. 3–7. Each of these chapters is dedicated to one of the major steps of the data matching process. They each present detailed descriptions of both traditional and state-of-the-art techniques, including recently proposed research approaches. Advantages and disadvantages of the various techniques are discussed. Each chapter ends with a section on practical aspects that are of relevance when data matching is employed in real-world applications, and with a section on open problems that can be the basis for future research.

Chapter 3 discusses the importance of data pre-processing (data cleaning and standardising), which often has to be applied to the input databases prior to data matching in order to achieve matched data of high quality. The topic of Chap. 4 is the different indexing (also known as blocking) techniques that are aimed at reducing the quadratic complexity of the naive process of pair-wise comparing each record from one database with all records in the other database. The actual comparison of records and their attribute (or field) values is then covered in Chap. 5, with an emphasis put on the various approximate string comparison techniques that have been developed. How to accurately classify the compared record pairs into matches and non-matches is then discussed in Chap. 6. Both supervised and unsupervised classification techniques, and pair-wise and collective techniques are presented. Finally, Chap. 7 describes how to properly evaluate the quality and complexity of a data matching exercise. This chapter also covers the manual clerical review process that traditionally has been (and commonly still is) used within certain data matching systems, and the various publicly available test data collections and data generators that can be of value to both researchers and practitioners.

The final part of the book then covers additional topics, starting in Chap. 8 with a discussion of the privacy aspects of data matching, which can be of importance because personal information is commonly required for matching data. This chapter also provides an overview of recent work into privacy-preserving data matching (how databases can be matched without any private or confidential information being revealed). Chapter 9 presents a series of topics that can be of interest to both practitioners as well as the data matching research community. These topics include matching geo-spatial data, matching unstructured or complex types of data, matching data in real-time, matching dynamic databases, and conducting data matching on parallel and distributed computing platforms. This chapter also includes a list of open research topics. Finally, the book concludes in

Chap. 10 with a checklist of how data matching systems can be evaluated, and a brief overview of several freely available data matching systems.

Rather than providing definitions of relevant terms and concepts throughout the book, a glossary is provided at the end of the book (on page 243 onwards) that can help the reader to access the terms and concepts they are unfamiliar with.

Intended Audience

The aim of this book is to be accessible to researchers, graduate and research students, and to practitioners who work in data matching and related areas. It is assumed the reader has some expertise in algorithms and data structures, and database technologies. Most chapters of this book end with a section that provides pointers to further background and research material, which will allow the interested reader to cover gaps in their knowledge and explore a specific topic in more depth.

This book provides the reader with a broad range of data matching concepts and techniques, touching on all aspects of the data matching process. A wide range of research in data matching is covered, and critical comparisons between state-of-the-art approaches are provided. This book can thus help researchers from related fields (such as databases, data mining, machine learning, knowledge engineering, information retrieval, information systems, or health informatics), as well as students who are interested to enter this field of research, to become familiar with recent research developments and identify open research challenges in data matching. Each of the Chaps. 3–9 contain a section that discusses open research topics.

This book can help practitioners to better understand the current state-of-the-art in data matching techniques and concepts. Given that in many application domains it is not feasible to simply use or implement an existing off-the-shelf data matching system without substantial adaption and customisation, it is crucial for practitioners to understand the internal workings and limitations of such systems. Practical considerations are discussed in Chaps. 3–8 for each of the major steps of the data matching process.

The technical level of this book also makes it accessible to students taking advanced undergraduate and graduate level courses on data matching or data quality. While such courses are currently rare, with the ongoing challenges that the areas of data quality and data integration pose in many organizations in both the public and private sectors, there is a demand worldwide for graduates with skills and expertise in these areas. It is hoped that this book can help to address this demand.

Acknowledgments

I would like to start by thanking Tim Churches from the New South Wales Department of Health and Sax Institute, for highlighting in 2001 to me and my colleagues at the Australian National University that the area of data matching can provide exciting research opportunities, and for supporting our research through funding over several years. Without Tim, much of the outcomes we have accomplished over the past decade, such as the FEBRL data matching system, would not have been possible. Thanks goes also to Ross Gayler and Veda Advantage, David Hawking and Funnelback Pty. Ltd., and Fujitsu Laboratories (Japan). Without their support we would not have been able to continue our research in this area. I also like to acknowledge the funding we received for our research from the Australian Research Council (ARC) under two Linkage Projects (LP0453463 and LP100200079), and from the Australian Partnership for Advanced Computing (APAC).

Along the way, I received advice from experienced data matching practitioners, including William Winkler and John Bass, who emphasized the gap between data matching research and its practical application in the real world. A big thanks goes also to all my students who contributed to our research efforts over the years: Justin Xi Zhu, Puthick Hok, Daniel Belacic, Yinghua Zheng, Xiaoyu Huang, Agus Pudjijono, Irwan Krisna, Karl Goiser, Dinusha Vatsalan, and Zhichun (Sally) Fu.

Large portions of this book were written while I was on sabbatical in 2011, and I would like to thank Henry Gardner, Director Research School of Computer Science at the Australian National University, for facilitating this relief from my normal academic duties. My colleagues Paul Thomas and Richard Jones have provided valuable feedback on early versions of this book, and I would like to thank them for their efforts. Insightful comments by William Winkler, Warwick Graco, and Vassilios Verykios helped to clarify certain aspects of the manuscript.

The list of research challenges and directions provided in Sect. 9.6 was compiled with contributions from Brad Malin, Vassilios Verykios, Hector Garcia-Molina, Steven (Euijong) Whang, Warwick Graco, and William Winkler (who gave the striking comment that "if one goes back 50 + years, these five issues were present" regarding the major challenges of data matching from the perspective of an experienced practitioner).

I would also like to thank the two anonymous reviewers who provided valuable detailed feedback and helpful suggestions. The task of proof-reading of the final manuscript was made easier through the help of my colleagues and students Paul Thomas, Qing Wang, Huizhi (Elly) Liang, Banda Ramadan, Dinusha Vatsalan, Zhichun (Sally) Fu, Felicity Splatt, and Brett Romero, who all detected the small hidden mistakes I had missed.

I also like to thank the editors of this book series, Mike Carey and Stefano Ceri, and to Ralf Gestner from Springer, who all supported this book project right from the start.

And finally, last but not least, a very big thanks goes to Gail for her love, encouragement and understanding.

Canberra, 29 April 2012   Peter Christen

Contents

Part I
Overview

Chapter 1
Introduction

1.1 Aims and Challenges of Data Matching

Given the ever-increasing amount of data that are being collected, not just by businesses and government organisations but increasingly also by individuals, the past decade has seen strong interest in novel techniques that allow the efficient processing, management and analysis of large data collections. The fields of *data warehousing* and *data mining* have gained immense interest in both academia and industry. While data warehousing is concerned with the efficient processing, integration and storage of large amounts of data into clean, consistent and persistent forms that enable basic statistical analysis, data mining is aimed at discovering new and potentially valuable information from such large data collections [135].

As businesses, public bodies and government agencies are drowning in an ever-increasing deluge of data, the ability to analyse their data in a timely fashion can provide a competitive edge to a commercial enterprise, lead to improved productivity for government agencies and be of vital importance to national security. In many large-scale information systems and data mining projects, data from multiple sources need to be integrated and matched in order to improve data quality, enrich existing data sources or facilitate data mining that is not feasible on a single database. The analysis of data integrated from disparate sources, either within an organisation or between different organisations, can lead to much improved benefits compared to analysing databases in isolation. Integrated data can also allow types of data analyses that are not feasible on individual databases, such as the identification of adverse drug reactions in particular patient groups, or the detection of terrorism suspects through the analysis of certain suspicious patterns of activities [44, 58, 103, 143].

Integrating data from different sources consists of three tasks. The first task is *schema matching* [224], which is concerned with identifying database tables, attributes and conceptual structures (such as ontologies, XML schemas and UML diagrams) from disparate databases that contain data that correspond to the same type of information. The second task, the topic of this book, is *data matching*, the task of identifying and matching individual records from disparate databases that refer to

P. Christen, *Data Matching*, Data-Centric Systems and Applications,
DOI: 10.1007/978-3-642-31164-2_1, © Springer-Verlag Berlin Heidelberg 2012

the same real-world entities or objects. A special case of data matching is *duplicate detection*, the task of identifying and matching records that refer to the same entities within a single database. The following Sect. 1.2 discusses how data matching fits into the overall data integration process. The third task, known as *data fusion* [38], is the process of merging pairs or groups of records that have been classified as matches (i.e. that are assumed to refer to the same entity) into a clean and consistent record that represents an entity. When applied on one database, this process is called *deduplication*.

The records considered in data matching and deduplication generally refer to real-world entities. The attribute values in these records are descriptions of the identifying details of these entities, such as their names, addresses and so on. It is assumed that these records are available already in a certain structured format, for example consisting of a name attribute, an address attribute, a date-of-birth attribute, etc. Data matching does not consider the extraction of entity information from unstructured documents (such as e-mails, news articles, police reports or scientific publications), or the scanning and optical character recognition (OCR) of names and addresses from letters and parcels. It is assumed that these *information extraction* [230] steps have already been conducted and that the records to be matched are stored in well-defined files or database tables.

Most commonly, the records to be matched across two or more databases, or to be deduplicated in a single database, correspond to people. They can, for example, refer to customers in a business database, employees in a company data warehouse, tax payers or welfare recipients in government databases, patients in hospital or private health insurance databases, known criminals and terrorism suspects in law enforcement and national security databases, or travellers in the databases held by airlines, and government departments of immigration and homeland security.

Besides people, other entities that sometimes have to be matched include records about businesses, publications and bibliographic citations, Web pages and Web search results or consumer products. In applications such as Web search and digital libraries, for example, it is important that duplicate documents (such as Web pages and bibliographic citations) in the results returned by a search engine are removed before the results are being presented to a user [131]. For automatic text indexing systems, it is important that duplicates are eliminated before the indexing takes place in order to reduce storage requirements and computational efforts [245].

With the increase in e-Commerce in recent years, another application where data matching has become of importance is comparative online shopping. Because consumer products in different online stores often have slightly different product descriptions (such as 'Canon PowerShot D10 Digital Camera' or 'Canon D10 12.1MP 3 × OPT ZOOMOIS Underwater Camera'), identifying which product description corresponds to which actual product can become difficult [33].

The task of identifying and matching records that refer to the same entities within one or across several databases is challenging for several reasons. The following sections highlight some of the major challenges. They will be further discussed in the relevant chapters later in this book.

1.1.1 Lack of Unique Entity Identifiers and Data Quality

Generally, the databases to be matched (or deduplicated) do not contain unique entity identifiers or keys. Examples of entity identifiers include unique patient or tax payer numbers, or consumer product codes. If such identifiers are available in all the databases to be matched, then the data matching task becomes a database join that can be implemented efficiently through SQL statements.

Even when entity identifiers are available in the databases to be matched, one must be absolutely confident in the accuracy, completeness, robustness and consistency over time of these identifiers, because any error in such an identifier will result in wrongly matched records.

As database owner, one must also be confident that there are no duplicate records in a database where different identifiers are used for the same entity. This situation is however common, for example in customer databases, where the same customer can have several records due to name variations or address changes.

If no entity identifiers are available in the databases to be matched, then the matching needs to rely upon the attributes that are common across the databases. If the databases contain information about people, then these common attributes can be names, addresses, dates of birth and other partially identifying personal details. The quality of such information can however be low, as personal details can be wrong, incomplete and they often change over time. Data matching based on such 'dirty' data is challenging, as will be discussed in Chaps. 3 and 5.

1.1.2 Computation Complexity

When matching two databases, potentially each record from one database needs to be compared with all records in the other database in order to determine if a pair of records corresponds to the same entity or not. When deduplicating a single database, each record potentially needs to be compared with all others. The computation complexity of data matching therefore grows quadratically as the databases to be matched get larger.

On the other hand, the number of potential true matches (i.e. pairs or groups of records that refer to the same entity) only grows linearly with the size of the databases to be matched. If it is assumed that the databases to be matched do not contain duplicate records, then the maximum possible number of true matches is limited by the size of the smaller of the two databases.

This computational challenge is addressed by techniques that aim to efficiently and effectively remove record pairs that likely do not refer to matches, while selecting candidate record pairs for detailed comparison and classification that likely will be matches. This topic is covered in detail in Chap. 4.

1.1.3 Lack of Training Data Containing the True Match Status

In many data matching applications, the true status of two records that are matched across two databases is not known, i.e. there is no ground-truth or 'gold standard' data available that specifies if two records correspond to the same entity or not. This is different from many other data mining or machine learning applications where training data are readily available. Without extra information (such as contacting individuals and asking for the correctness of their personal details) one cannot be sure that the outcomes of a data matching project are correct. This is especially a problem for large databases that cover large portions of a population, as is the case in health or government data collections. In such applications, accurately assessing data matching quality and completeness is challenging. Chapters 6 and 7 discuss this challenge in depth.

1.1.4 Privacy and Confidentiality

As previously mentioned, with data matching commonly relying on personal information such as names, addresses and dates of birth of individuals, privacy and confidentiality need to be carefully considered. This is especially the case when databases are matched between organisations, or when the outcomes (the matched data set) are to be used by an external organisation or by individuals such as academic researchers. The analysis of matched data has the potential to uncover aspects of individuals or groups of entities that are not obvious when a single database is analysed separately.

For example, the outcomes of analysing matched health and population databases can potentially lead to discrimination against certain groups of individuals, if it is discovered that these people have a higher risk of getting a certain serious or infectious disease. The discrimination could be in the form of higher life insurance premiums, or even that these individuals would find it much harder to gain employment due to their potentially increased risk of long-term illness.

In recent years, research into the development of techniques that facilitate *privacy-preserving record linkage* has received attention from areas such as health informatics and data mining. The aim of this research is to facilitate the matching of data across organisations without compromising the privacy and confidentiality of the data to be matched. Chapter 8 provides an overview of this challenging topic.

1.2 Data Integration and Link Analysis

Data matching is a commonly required step in the much larger process of *data integration* [178]. While data matching is concerned with identifying and matching individual records that refer to the same entities from disparate databases, data inte-

PatTbl

PatientID	Name	DOB	Age	Gender	StreetAddress	Suburb	Postcode
P1273489	John Smith	8/10/1960	51	M	8/42 Miller Street	Melbourne	3011
Q6549234-2	Mick Meyer	30/01/1948	63	M	10 Port Road	Ferny Grove	7004
P7693427-8	Joanna Smith	12/11/1984	27	F	76 George Crest	Sydeny	2020

AdmittedPatients

PID	Surname	GivenName	BirthDate	Sex	AID
25198	Smith	Jo Anna	19841112	1	A347
55642	Smith	John W.	19601008	0	A135
15907	Meier	Michael	19480101	0	A810
99801	Meyer	Mike	19790320	0	A135

Addresses

AID	Street	Location
A135	42 Miller St	3000 Melbourne
A347	16 George Crs	2000 Sydney
A810	PO Box 553	7000 Brisbane

Fig. 1.1 Three example hospital database tables that illustrate commonly occurring challenges with data integration. It is assumed that the **PatTbl** table originates from one database, while the two tables **AdmittedPatients** and **Addresses** are sourced from a second database. The **AID** is a unique identifier (key) for each address that links the **AdmittedPatients** table with the **Addresses** table

gration is the overall process of integrating heterogeneous databases, data warehouses or data repositories to provide a unified view of the available data. This process is highly significant, for example, for company mergers, collaborative e-Commerce projects, data mash-ups and scientific collaborations [24].

Data stored in disparate databases are usually heterogeneous not only at the record (instance) level, but also at the structural database (table) level. As illustrated in Fig. 1.1, the possible differences in the way data are represented and stored include [224]:

- Tables that contain the same type of information can have different names, and one table in one database can contain the same (or similar) type of information as several tables in another database. In the given example, the **PatTbl** table contains the details of hospital patients, as do the two tables **AdmittedPatients** and **Addresses**. Detecting which table corresponds to which (combination of) other table(s) is challenging if the number of tables in two databases is large.
- Attributes that contain the same type of information can have different names, and even if they contain the same information their content can be formatted or encoded differently. In the given example, the 'DOB' and 'BirthDate' attributes follow different date conventions, the first being 'DD/MM/YYYY' and the second being 'YYYYMMDD', with 'DD' representing day numbers, 'MM' month numbers and 'YYYY' year numbers. The 'Gender' attribute in the **PatTbl** table contains the same information as the 'Sex' attribute in the **AdmittedPatients** table, but they are encoded differently, with 'M' in the **PatTbl** table corresponding to '0' in the **AdmittedPatients** table, and 'F' to '1'.
- Attributes that contain compound information in one database can be split into several attributes in another database, such as the 'Location' attribute from the **Addresses** table which corresponds to the combined 'Suburb' and 'Postcode' attributes in the **AdmittedPatients** table.

- Within one database, some attributes can be derived from other attributes that are in different tables. For example, values in the 'Age' attribute in the **PatTbl** table were calculated in 2011 based on the corresponding 'DOB' values in the same table.
- Attribute values can also be recorded using different measurements, for example dollar amounts in a multinational business database will potentially correspond to national currencies such as US, Australian or New Zealand dollars.
- Entity identifiers, such as the shown 'PatientID' and 'PID' attributes, are potentially not the same across two or more databases. Exemptions are national identifiers such as social security, drivers licence or tax file numbers. In many countries, however, the use of such identifiers for data integration is strictly regulated or even prohibited.

As these examples show, data integration has many challenges, both at the database structure and the record level. When matching schemas, a careful analysis of the names, types of content and other meta-data of the available attributes can help to identify which attributes correspond to each other [24, 224]. Meta-data can include information such as the number of different values in an attribute and their frequency distribution, descriptions of the sources of the attribute values, or their structure and encoding (such as references to external encoding dictionaries).

Data integration should however not consider schema matching and data matching independently, but rather in an integral fashion [299]. Commonly, data integration is a semi-automated process, where human insight can provide initial information about which database tables potentially correspond to each other. Such hints can then be used to bootstrap a tool-supported integration process.

The content of attributes can help in the schema matching process. Correlation analysis can help detect attributes that contain related information. Data matching across two databases can identify records that correspond to the same entities, which in turn can help identify which attributes correspond to each other [299]. For example, if an approximate string matching algorithm (as will be discussed in Chap. 5) is applied on the two attributes 'StreetAddress' and 'Address' in Fig. 1.1, the correspondence of these two attributes can be inferred based on their similar content. A process that exploits information both at the schema and the record level, and iteratively refines the integration process, can therefore lead to efficient and accurate data integration.

Once databases are matched at the schema level, individual records need to be matched to identify which records in two or more databases correspond to the same real-world entities. The remainder of this book covers this second data integration task in detail.

The final task in the data integration process, after pairs or groups of records have been identified that refer to the same entities, is to consolidate and merge matching records into a single consistent and clean representation for each entity. This step is also known as data fusion [38]. The major challenge of data fusion is how conflicts are resolved when the records that correspond to one entity contain different attribute values. Different aggregation functions can be applied, such as taking the minimum, the maximum, the oldest or the newest of a set of different attribute values, calculating an average of values (if feasible), or taking the value from a data source that is more trustworthy than another. Bleiholder and Naumann [38] provide an excellent survey of data fusion and its challenges, and how fusion can be achieved within a database framework using relational operators. Their survey also includes an overview of data fusion systems.

Besides relational databases, recent interest in other structured types of data, such as XML schemas and ontologies, has sparked the development of novel techniques that are aimed at integrating and matching such types of data [24, 120, 270]. These techniques commonly combine both semantic and syntactic similarities into one overall similarity value which is used to detect similar entities. The outcomes of such matching can, for example, help to detect ontologies that are variations of others, or XML schemas that are parts of other, larger schemas.

Data matching is also an integral part of link analysis (or link mining or network analysis) [296], techniques that are concerned with the exploration, modelling, evaluation and prediction of connections between nodes in networks. In application areas such as counter-terrorism, crime investigations and fraud detection, the nodes (entities) in such networks commonly correspond to people, and the connections (relationships) between these nodes refer to their interactions (such as phone calls, financial transactions, shared addresses, shared activities and so on). Link analysis often involves the interactive visualisation of a sub-graph of a network that is of interest, for example for a criminal investigation. Data matching is an important component of link analysis because it allows nodes to be identified that correspond to the same real-world entities even though they have different attribute values.

1.3 A Short History of Data Matching

Data matching has a long history. Even before the invention of modern computers, statisticians and public health researchers were interested in identifying records that belonged to the same entity from a single or several disparate databases.

In 1946, Dunn used the term *record linkage* [97] to describe the idea of assembling a *book of life* for every individual in the world. Each of these books would start with a birth record and end with a death record, and in between it would consist of records about an individual's contacts with the health and social security systems, and also include marriage and divorce records. Dunn realised that having such books of life for all individuals in a population will provide a wealth of information that allows governments to improve national statistics, better plan services and also improve the identification of individuals. Dunn also recognised the difficulty of dealing with data quality issues, such as common names, errors and variations in the data.

In the 1950s and early 1960s, Howard Newcombe et al. [197, 198] then proposed the use of computers to automate the data matching process. He also developed the basic ideas of the successful *probabilistic record linkage* approach. In his approach, the phonetic Soundex [57, 201] encoding is applied to attributes such as surnames to overcome name variations. Based on the distribution of attribute values, match and non-match weights (also called agreement and disagreement weights) are calculated and used to decide if two records correspond to a match or not.

Based on Newcombe's ideas, in 1969 the two statisticians Ivan Fellegi and Alan Sunter published their seminal paper on probabilistic record linkage [108]. Their theory proved that an optimal probabilistic decision rule can be found under the assumption that the attributes used in the comparison of records are independent of each other. This pioneering work has been the basis for many data matching systems and software products, and it is still widely used today. The probabilistic record linkage approach will be covered in detail in Chap. 6. Okner in 1974 summarised several data matching projects in the taxation domain [203], and he described the difficulties encountered with exact matching approaches.

Over the past few decades, the basic probabilistic approach has been extended in various ways. Notable work was conducted by William Winkler and his colleagues at the US Census Bureau in the 1990s [286]. They developed techniques that allowed for variations in strings through the use of approximate string comparison functions [215], used frequency-based match and non-match weights, and employed the expectation–maximisation (EM) algorithm [280] to improve the estimates of the matching parameters required in the probabilistic record linkage approach.

Concurrently to the work done by statisticians and public health researchers, the database community has developed techniques to find duplicate records in a single database [140], and to improve the quality of databases as part of the data cleaning process [224]. Duplicate records commonly occur in customer databases, and as will be described in Sect. 1.4.5 below, identifying and removing duplicates is an important task. The work conducted by database researchers was not based on the probabilistic approach developed by Fellegi and Sunter. Rather, they employed sorting of the databases according to the attributes to be compared in order to bring similar records together [140, 141], and used string comparison functions to detect approximate similarities [190, 191]. It is interesting to note that until the late 1990s few cross-references could be found between computer science and statistical or health publications in the area of data matching.

The last 10 years have seen a strong increase in data matching research in the computer science domain, especially in areas like data mining, machine learning and information retrieval, as well as the database and data warehousing communities [103]. As larger databases have been collected by many organisations, and data quality has been recognised as a major challenge to utilising these data [19, 177], the task of identifying records that refer to the same entities in disparate databases has become more pervasive than ever. In the past few years, novel techniques have been developed that employ sophisticated machine learning, natural language processing and graph-based approaches in order to improve both data matching quality and enable the matching of very large databases that contain many millions of records. Winkler in 2006 provided an excellent overview of work done both in the computer science and statistical research domains [284].

A different avenue of data matching research in the past decade has been the development of techniques that allow the matching of databases without the need of any private or confidential data to be exchanged between the organisations that conduct the matching. These techniques, known as *privacy-preserving record linkage*, will be described in Chap. 8.

Because the task of data matching has been investigated independently in different research domains, it has been given a variety of names. While health researchers and statisticians speak of record or data linkage, computer scientists name the same process as data, record or object matching, entity resolution, co-reference resolution, object identification, data reconciliation, citation or reference matching (when applied to bibliographic databases [185]), identity uncertainty, duplicate detection or deduplication (when applied to one database only), authority control, or approximate string join in the context of databases. In the business-oriented processing of databases and data-warehousing, data matching is also known as the merge/purge problem, list washing, data cleansing or field scrubbing, and is seen as one crucial step of the overall extraction, transformation and loading (ETL) process [224].

Today, there are dozens, if not hundreds, of commercial data matching and deduplication products and solutions on the market. Many of these are either stand-alone packages that are specialised for a certain type of application, such as the deduplication of mailing lists or the matching of health databases; or data matching is one component of a much larger business intelligence, data integration, data quality or customer relationship management system. Chapter 10 will further discuss commercial as well as freely available data matching systems, and also provide a checklist that can be used to evaluate the requirements of data matching systems. Most modern commercial database systems also contain some functionality that can be used for data matching, such as phonetic encoding and approximate string comparison functions. These techniques will be presented in detail in Sect. 4.3 and Chap. 5, respectively.

1.4 Example Application Areas

The following sections describe several example application areas where data matching is an important component of larger information systems, of government and business processes, or of research endeavours. For each area, the unique aspects and challenges encountered are discussed.

1.4.1 National Census

National census agencies around the world collect data about various aspects of the population, culture, economy and the environment in their respective country. This information is then collated into a diverse range of statistical reports that are used by governments and businesses to plan the allocation of funding and resources.

Data matching has been recognised as an important tool for census statistics. It allows the reuse of existing data sources to compile new statistical data sets, and thus reduces the costs and efforts required to conduct large-scale census collections [119]. It also helps to improve data quality and integrity, as matching data from different census collections can help detect and correct conflicting or missing information, or improve estimates of population sizes through capture–recapture techniques [282].

Data matching can also be used to generate longitudinal data sets, by matching census data that have been collected at different instants in time (for example every 5 or 10 years). It is commonly recognised that longitudinal data are an important source of information about how the characteristics of a population change over time. Different countries have different laws and regulations that govern what kind of data matching can be done. In Australia, for example, name and address details collected in national censuses need to be destroyed within 1 year after collection (both the physical paper-based census forms as well as any electronic versions of these data). Such restrictions make it very challenging to create longitudinal data sets, because the matching has to rely upon information such as age, gender, birthplace, religion, highest educational qualification and so on [86].

The US Census Bureau has been one of the early adopters of data matching. Not only has the Bureau applied existing data matching techniques on a regular basis, it has also been at the forefront of data matching research and development over several decades [215, 280, 286]. The size of the data collections the Bureau needs to match can be in the order of several hundred millions of records, and therefore it has also been developing large-scale and parallel data matching techniques [295]. The various techniques developed by the US Census Bureau will be described in the corresponding chapters later in this book.

1.4.2 The Health Sector

Besides national censuses, the health sector has been a second application area that has pioneered the use of data matching for several decades [97, 199]. Over the duration of a person's life, the detailed information collected by doctors, hospitals, heath insurers and pharmacies, results in a detailed picture of the health of an individual. On a population scale, this information can be of high value.

Matched health data allow the reuse of existing data sources for new studies, thereby reducing costs and efforts in data acquisition. For example, it enables the investigation of adverse drug reactions in certain patient groups. Matching patient addresses with spatial data can for example lead to the discovery of correlations between environmental factors and local hot-spots of disease cases.

In the UK, the *Oxford Record Linkage Study*, which started in the 1960s, was aimed at developing novel computer-based data matching techniques, and applying these techniques on birth, death and hospital data of around 350,000 individuals [119]. This allowed the study of associations between certain diseases, and using longitudinal matched data enabled the analysis of occupational mortality, migration and related socio-economic factors.

Arguably one of the most successful practical health data matching programs worldwide has been conducted in Western Australia since the mid-1990s. Based on a best-practice protocol of separating the personal identifiers required for matching from the medical information needed for research studies [161], data from various health (as well as non-health) sources have been matched, and a chain of records has been generated for each individual person identified. A recent publication has summarised over 700 outputs produced by this program from 1995 to 2003 [44]. Some of the significant outcomes include improvements in health policies (like regular physical examination for mental health patients) and changes to clinical practise (like installation of shock advisory defibrillators in all ambulances and hospital wards, or community-based services for psychiatric patients at risk of suicide).

Several other countries have implemented similar health data matching programs. However, the sensitivity of personal health records and privacy and confidentiality concerns have thus far limited the application of data matching in the health sector in many countries. Current research efforts aim to develop techniques that allow data matching while at the same time preserving the privacy of the data to be matched. Chapter 8 will discuss the topic of privacy within the domain of data matching in more detail.

1.4.3 National Security

After the terrorism attacks on the US in 2001, the US government (as well as the governments of other western countries) significantly increased their efforts and funding into advanced analytics programs, with the aim to better detect and even prevent future acts of terrorism. Compared to traditional armies, terrorists are much harder to identify and track, because they are loosely organised in secret networks and individual cells, they hide and strike infrequently, and they receive funding from a variety of sources [213]. Nevertheless, terrorists do leave transactions and records about their communications and activities in the online information space.

The objectives of counter-terrorism initiatives such as the *total information awareness* (TIA) and the *multistate anti-terrorism information exchange system* (MATRIX) programs [109] was to apply advanced techniques from domains including data matching, data and link mining, biometrics, natural language processing and image recognition, to detect unusual patterns within and across a variety of databases. The databases that were to be accessed in these programs originate from both government agencies as well as private organisations such as banks, airlines, flight schools and car hire companies.

The challenges of applying data matching in such applications are manifold. The databases to be matched are very large and contain many millions of records, and they are also very diverse. Different details about individuals are stored within these databases. The records of individuals are also likely collected at different points in time, therefore address and name details for the same person might differ. Criminal individuals and terrorists are also likely to use faked or modified identities in order to hide their activities [267].

Finding the very small number of potential terrorists out of a population of hundreds of millions of individuals is akin to finding a needle in a very large haystack. The accuracy of the data matching (and data mining) models need to be extremely high, as otherwise the number of (potentially false) positive matches (records from different databases that a model associates with the same individual) can become too large, which prevents effective investigations [153]. Furthermore, the identification of a suspect is often required in real-time by querying a large database with the identity details of the suspect. The topic of real-time data matching will be covered in more detail in Chap. 9.

It has been reported that by 2008, the number of individuals on US terrorism watch lists was nearly 500,000 [171]. Most of them are likely ordinary citizens that have some characteristics that make them suspects, or that at some time behaved in such ways that was significantly different from what is normally expected behaviour. Stories of innocent citizens being scrutinised each time they travelled via plane, and concerns by the public about government agencies matching and analysing their data and thus invading their privacy, lead to a backlash against programs such as TIA and MATRIX, which had to be abandoned or modified. The issue of privacy with regard to data matching will be further discussed in Chap. 8.

1.4.4 Crime and Fraud Detection and Prevention

Fighting crime and fraud today relies on sophisticated information systems that allow the accurate identification of individuals that are suspects in a crime or fraud investigation [54]. Data matching techniques are integral parts of such modern crime fighting information systems. Criminals commonly provide modified or even fictitious identifying personal details when questioned by law enforcement officers [267]. These deceptive identity details can for example be addresses of acquaintances, dates of birth of deceased persons, or faked social security or drivers license numbers.

A major challenge when applying data matching in the domain of crime and fraud detection is therefore that, unlike in most other domains, variations and errors do not just occur because of data entry errors and the changing nature of people's personal details, but because individuals deliberately modify their details because they do not want to be identified. These deliberate changes are done in such ways that the modified details look real (and possibly correspond to another individual), and like 'innocent' variations or errors that could have occurred by chance.

Data matching techniques can help to detect if these deceptive identity details do refer to a real person or not. Using databases of known criminals and their aliases allows law enforcement officers to identify an individual under questioning. To be effective, data matching techniques in this application area must facilitate the detection of potentially matching records in (near) real-time, i.e. within a few seconds.

How to accomplish such real-time data matching in large databases will be discussed in Chap. 9.

A similar application where data matching is increasingly being employed is identity verification aimed at reducing identity crimes [212]. Identity crimes are on the increase in many countries, resulting in losses of billions of dollars to financial organisations and grave social implications for the individuals concerned [10]. An identity crime occurs when a fraudster gains access to services and benefits by using a false identity. With the increasingly widespread use of electronic financial trans-actions and online government services it has become essential to verify the identity of absent participants with a high level of certainty. Based on a statistical analysis of 300 million accounts, it has recently been reported that around 90 % of success-fully opened fraudulent bank accounts in the US used synthetic identities [206], with almost 75 % of all dollars lost due to synthetic identity fraud.

Data matching is a crucial component to identity verification, as it allows match-ing of the identifying details provided by an individual with a variety of databases that contain verified and accurate entity records. Such databases can include voter registrations (also known as electoral rolls), drivers license and social security num-ber databases, and telephone directories. Matching records from these databases will allow an overall picture to be built of the individual under consideration, and a risk score can be calculated that provides an indicator of the likelihood that the presented identity details do refer to a real person or not. Similar to data matching used in fight-ing crimes, real-time matching of a stream of individual query records on several large databases is required.

1.4.5 Business Mailing Lists

Many businesses spend significant amounts of resources and money on advertising their products and services. They sometimes base their advertisement campaigns, such as mailing out flyers (also known as handouts, leaflets, circulars or pamphlets), on their databases of existing customers. The quality of the data a business has collected about their customers is therefore of significant value. As can be seen from the examples shown in Fig. 1.2, having several records about the same customer results in money being wasted on mailing several copies of an advertisement flyer to one individual.

There are several challenges businesses face when trying to maintain customer databases of high quality. First of all, address details change when people move, and when somebody gets married or divorced they might change their surname. This can easily result in duplicate records about the same individual. Second, many customers do not care if there are several duplicates about them in a business database, as long as the products or services they bought from a business are being delivered. A customer is also unlikely to complain when they receive several copies of the same advertisement flyer.

Fig. 1.2 Three real, scanned
address duplicates the author
received some years back in
the form of three copies of
an advertisement flyer by the
same company

DR PETER CHRISTEN
DEPARTMENT OF COMPUTER SCIENCE
AUSTRALIAN NATIONAL UNIVERSITY
CSIT BUILDING (108) NORTH ROA
CANBERRA ACT 2600

PETER CHRISTEN
AUSTRALIAN NATIONAL UNIVERSITY
DEPT OF COMPUTER SCIENCES
BARRY DRIVE BLDG 108
CANBERRA ACT 0200

Dr PETER CHRISTEN
DEPARTMENT OF COMPUTER SCIENCE
AUSTRALIAN NATIONAL UNIVERSITY
CANBERRA ACT 0200

A third challenge occurs when two or more businesses collaborate with each
other, for example on cross-marketing campaigns, and therefore customer records
from different sources need to be matched in order to build a combined database of
the individuals to whom advertisements shall be sent out. It will be likely that the
customer information from the different sources will have different formats, different
types of information will be available, and individual records will have been collected
at different points in time. This can make it difficult to identify if two or more records
correspond to the same customer or not.

The final challenge of data quality can be addressed by businesses themselves. As
more and more customers shop online, it is crucial to assure the information entered
by customers on Web forms is accurate, complete, and validated as much as possible.
The process of validating input values can include the verification of addresses and
their comparisons with a database of existing street addresses. The postal services
in many countries now provide databases that contain detailed street addresses that
allow checking if an entered address is valid, and if not alert the person entering the
data (either the customer or an employee of a business). Furthermore, addresses that
are segmented into specific input fields each containing a single piece of information,
rather than free-format text input fields, will lead to much improved data quality. Not
limiting the length of the data entered into an input field is another way to improve
data quality, as otherwise customers will be forced to abbreviate long names. This
will result in unusual and unstandardised values. When data are entered, they should
also be matched in real-time with records in a database of existing customers to
prevent duplicate records being created in the first place. This matching process can
be based on a known name and address combination, or a known telephone number,
or ideally a known customer identifier. The topics of data quality and of data entry
errors will be discussed in detail in Chap. 3.

1.4.6 Bibliographic Databases

Research in many domains is increasingly being disseminated electronically and through online databases such as *Springer Link*,[1] *Elsevier Scopus*,[2] *Thompson Web of Knowledge*,[3] the *ACM Digital Library*,[4] *IEEE Xplore*,[5] or *Google Scholar*.[6] These databases allow researchers to access millions of publications. They also provide services such as citation and impact analyses, alert services for new publications by an author, and notifications of new citations for given publications.

While such bibliographic databases facilitate a much faster dissemination of new knowledge, government funding agencies around the world also increasingly rely upon these databases to calculate numerical metrics to assess the impact and significance of individual researchers, research groups, faculties and even institutions. These metrics are then used to allocate funding to individual research projects or institutions, as well as to support decision making within academic promotion processes. Measures such as the *h-index*[7] and its variations [144] calculate a single numerical value for an individual researcher based on the number of citations their publications have attracted. The quality of the data used in such evaluations need to be high, otherwise these numerical impact measures will become questionable.

There are several challenges involved in creating and maintaining bibliographic databases [176, 183]. One challenge is their size, with some of the larger of these databases containing well above 25 million publications. However, the biggest challenge is that it is very common to have several researchers with the same surname and the same initials in a database, some even working in the same research domain. Even if full given names are provided, it is often not clear if two publications were written by the same individual or not. Additionally, journal and conference names are commonly abbreviated, not following a standardised format, and therefore different variations of the same reference can often be found. Figure 1.3 shows an example of three variations of a reference to the same journal article taken from the Cora collection of machine learning publications [184]. In the domain of library research, data matching is also known as authority control [183].

On the other hand, bibliographic data have characteristics that make them very attractive for data matching research. Unlike other types of data that contain personal information, there are generally no privacy or confidentiality concerns when publication data are being matched, because such data are already publicly available. Additionally, such data contain information about several types of entities, including authors, publication venues (journals, conferences and workshops) and affiliations

---

[1] http://www.springerlink.com.

[2] http://www.scopus.com.

[3] http://apps.isiknowledge.com.

[4] http://portal.acm.org.

[5] http://http://ieeexplore.ieee.org.

[6] http://scholar.google.com.

[7] A researcher has an h-index of x if they have x publications that have each received at least x citations.

| Kearns, M., Li, M., and Valiant, L: Learning boolean formulas, ACM 41 (1994a), 1298-1328 |
| M. Kearns, M. Li and L. Valiant: Learning boolean formulas. Journal of the ACM, 41, 1994; 1298-1328. |
| M. Kearns, M. Li, and L. Valiant: Learning boolean formulae. Journal of the Association for Computing Machinery 41(6), 1995, pp. 1298–1328 |

Fig. 1.3 Three bibliographic records of the same machine learning publication, taken from the Cora data set [184]

(the institutions where authors work). This information can be used to build a graph of potentially matching entities, and evidence between one type of matching entity can help to infer the matching across another type of entity. For example, it is more likely that two authors with the same surname that work at the same university are the same individual compared to two authors who work at different universities. This type of information has been exploited by recent research into so-called *collective entity resolution*. These techniques are based on relational clustering or graph techniques, and they aim to find an overall optimal matching solution over a database of entities [31, 155]. These techniques will be discussed in detail in Chap. 6.

As a cautionary note, however, it is important to understand that data matching algorithms that were developed for the domain of bibliographic data should not be directly used for the matching of databases that contain personal information. The difference in the structure and content of the data in these two domains will require a careful assessment of how algorithms developed in one domain can be employed in another domain.

1.4.7 Online Shopping

As consumers worldwide tend to increasingly use online shopping for consumer products and services, Web sites that provide price comparisons have become popular. These sites allow consumers to either query a certain product or browse product categories, types and brands. A challenge for such comparison shopping sites is to identify which product descriptions in different online stores do correspond to the same item [33]. While certain types of items have unique identifiers, such as ISBN numbers for books, and systems such as electronic product Code (EPC) that are based on Radio Frequency Identification (RFID) codes are becoming more widespread, the descriptions of the same product are often quite different across several online stores. Figure 1.4 shows four example descriptions and product item numbers of the same digital camera taken from four different online stores.

To allow accurate and comprehensive price comparisons for an individual consumer product, a comparison shopping site needs to be able to accurately identify which product descriptions refer to the same actual product. Compared to other types of data that are commonly used for data matching (like name and address details of

Canon PowerShot G11 10 MP Compact Camera - 6.10mm-30.50mm	Item # 927909
Canon – PowerShot G-11 10.0 Mega pixel	Item# CANPSG11
PowerShot G11 Point & Shoot Digital Camera	Canon 3632B001
Canon PowerShot G11 10 Megapixel Compact Camera	MFG #: 3632B001

Fig. 1.4 Four descriptions of the same digital camera taken from different online stores. Not only are the descriptions different but also the (internal) product item numbers used

people), the variations in the descriptions of consumer products require different similarity measures. For example, the string similarities between the four descriptions shown in Fig. 1.4 are quite obvious. On the other hand, the differences between a description of this camera and its predecessor ('Canon PowerShot G10') might only be the single difference between the '0' and '1' digit in the camera's number. The number of megapixels (10 for this camera) complicates the similarity calculations for this example even further.

As this example highlights, data matching is often a very data dependent activity, and techniques such as approximate comparison functions need to be specifically designed for a certain task and data at hand. A wide range of such comparison functions will be described in Chap. 5.

1.4.8 Social Sciences and Genealogy

In recent years there has been a shift in the way social science research is being conducted. While traditionally small-scale surveys were the basis of most research studies in areas such as social welfare, political studies or demography, large population-based data collections are increasingly being used. One advantage of such large data collections is that they normally do not contain any bias that might have been introduced by a small survey data collection. A second advantage of large population databases is that they are being collected by government agencies as well as private sector organisations for purposes other than social science-oriented data analysis, and therefore often no costs are incurred by researchers.

In many cases, however, data from different sources need to be matched to allow certain studies. Matching demographic and health data, for example, can be used to analyse, among other things, epidemiological aspects of inherited disease patterns [121], and trends in nutritional states, health, mortality and the processes of ageing [111].

An important source of information for social scientists and lay people as well is historical census data. Population census returns provide invaluable snapshots of the state of a nation, and are the basis of modern public policy making. They help researchers understand how our ancestors lived, and the social and demographic changes in their societies. For individuals, online genealogical databases, either provided by national census agencies or private companies such as http://Ancestry. com, are a fascinating source of information that allows them to discover where their ancestors came from. Such online genealogical databases are now a billion-dollar industry.

Invaluable as they are, census returns are still only snapshots of moments in time, and even online genealogical databases commonly require the manual exploration and matching of census records across time. Matching records (that refer to the same individuals, families or households) from several census returns across time can greatly enhance their value [114–116]. Such matched data will provide social scientists with new insights into the dynamic character of social, economic and demographic changes [222, 227]. It will allow researchers to reconstruct the key life course events of large numbers of individuals, households and families, and ask new questions about changes in society and its history at levels of detail far beyond the scope of traditional methods of historical research. Such data will even facilitate epidemiological studies of the genetic factors of diseases such as cancer, diabetes or mental illnesses [121].

Figure 1.5 shows an original historical census return from the 1900 US Census. Matching such historical census data across time has many challenges. Data quality (scanning, optical character recognition (OCR), transliteration and data entry errors) is one obvious issue. Other problems occur due to the facts that historical addresses were not detailed, not standardised and they changed significantly over time as street names and numbers became more formalised. The frequency distributions of both family and given names were very clustered in the nineteenth century. It was, for example, not uncommon in the mid-nineteenth century in England that more than 10 % of the male population had the given name 'John', and more than 10 % of the female population had the given name 'Mary'. The discriminating power of name and address values, when used in matching of such historical data collections, is therefore quite different from their use in modern data collections.

1.5 Further Reading

Only a few books cover data matching in detail. Talburt [249] presents this topic as being part of the wider issue of information quality. He focuses on algebraic solutions and presents several systems, both commercial and open source. Similarly, Batini and Scannapieco [19] cover data matching as part of their book on data quality. A declarative approach to data matching was presented by Chan et al. [51], who also discuss semantic data matching.

A basic introduction to duplicate detection was recently provided by Naumann and Herschel [195], while the more traditional, statistical approach to data matching is covered in detail by Herzog et al. [143]. Older books and reports on this topic include Newcombe's handbook on record linkage [199] and Gill's report on using data matching for national statistics [119].

Two excellent survey articles on data matching have recently been written by Elmagarmid et al. [103], and Winkler [284], respectively. Bleiholder and Naumann

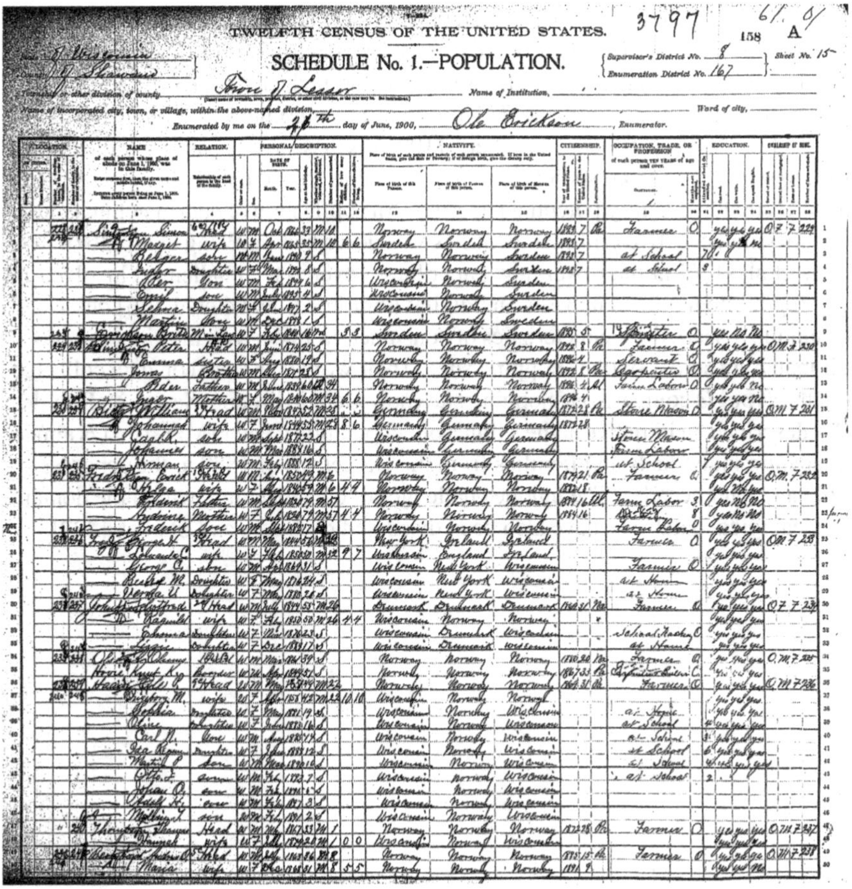

Fig. 1.5 An original historical census return form from the early twentieth century (sourced from http://familypedia.wikia.com). Each row contains the handwritten details of an individual, including their name, relation to head of household, gender, place of birth and occupation and education details. The left side contains location information in the form of a unique number for each household. Such census forms are the basis of modern genealogical databases. They are scanned, and often data are entered manually. It is easy to see that data quality is a major issue when matching such historical census data

provide an overview of data fusion and its application within relational database frameworks [38].

Winkler and his colleagues from the US Census Bureau over the past few decades have published a series of reports on using data matching for application within national censuses [149, 215, 280, 284, 286, 295]. Within the biomedical research literature, there is a very large body of studies that have been based on matched health data sets. Unfortunately, not many of these provide much details about the data matching techniques employed. Two recent comparisons of data matching systems used in the health area are by Gomatam et al. [124], who compared a deterministic with a commercial probabilistic system, and Campbell et al. [47], who evaluated three data matching systems that are used in the biomedical domain.

Some technical aspects of the *TIA* program are presented by Poindexter et al. [213], while Jonas [153] and Fienberg [109] discuss privacy and confidentiality issues that can arise from such programs, as well as the actual challenges of being able to accurately identify potential terrorists. An application of data matching aimed at identifying deceptive criminal identities was described by Wang et al. [267].

Lee et al. [176] in their overview paper have covered the challenges involved in matching bibliographic databases and digital libraries. Bilenko et al. [33] described a technique to learn the characteristics of consumer products with the aim to improve their matching across online shopping sites. Their work was based on data collected from the *Froogle* comparison shopping site (now Google Product Search[8]). Ruggles [227] described an initiative at the Minnesota Population Center (MPC) that is aimed at matching individuals and families from the US census of 1880 to the censuses of earlier and later decades. The unique challenges of data matching within the domain of genealogical databases have also been discussed by Quass and Starkey [222].

---

[8] http://www.google.com/shopping.

Chapter 2
The Data Matching Process

2.1 Overview

An overview of the data matching process with its five major steps is shown in
Fig. 2.1. The first step is the process of *data pre-processing*, which assures the data
from both sources are in the same format. The second step, *indexing*, aims to reduce
the quadratic complexity of the data matching process through the use of data structures that facilitate the efficient and effective generation of candidate record pairs
that likely correspond to matches (i.e. refer to the same real-world entity).

In the third step, the actual *record pair comparison* occurs, where candidate record
pairs are generated from the indexing data structures built in the previous step. These
pairs are compared using a variety of field and record comparison functions. In the
classification step, candidate record pairs are classified into *matches*, *non-matches*,
and *potential matches* (depending upon the decision model used [129]). If record
pairs are classified into potential matches, a manual *clerical review* process is needed
to decide their final match status (match or non-match). In the *evaluation* step, the
quality and completeness of the matched data, and the complexity of a data matching
exercise, are evaluated.

For the deduplication of a single database, all steps of the data matching process
are still applicable. Data pre-processing is important to assure the complete database
is in a standardised format. This is especially important if records have been added
to a database over time, potentially with changes in data entry techniques or methods
that lead to different data formats and encodings over time. The indexing step is also
of importance for deduplication, because comparing each record in a database with
all others has a quadratic computation complexity.

2.1.1 A Small Data Matching Example

To illustrate the various challenges and tasks involved throughout the data matching
process, an example consisting of two small database tables is used throughout this

P. Christen, *Data Matching*, Data-Centric Systems and Applications,
DOI: 10.1007/978-3-642-31164-2_2, © Springer-Verlag Berlin Heidelberg 2012

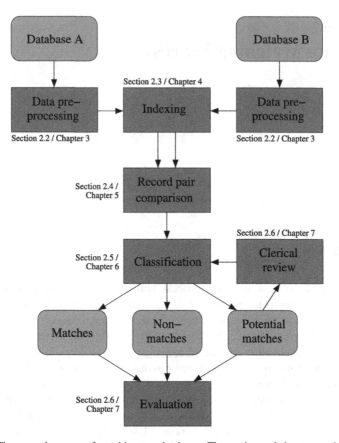

Fig. 2.1 The general process of matching two databases. The section and chapter numbers shown provide the road map through this chapter and Part II of this book

chapter. Figure 2.2 shows the two raw database tables that are to be matched. As can be seen, while they both contain name, address, and date of birth information, the structure of the two tables is different, as is the format of the values stored in the two tables. Each record is identified through a unique value in the 'RecID' attribute.

2.2 Data Pre-Processing

As the database tables in Fig. 2.2 show, data that are used for data matching can vary in format, structure and content. Because data matching commonly relies on personal information, such as names, addresses, and dates of birth, it is important to make sure that data sourced from different databases have been appropriately cleaned and standardised. The aim of this process is to ensure that the attributes used for the

Database A

RecID	Surname	GivenName	Street	Suburb	Postcode	State	DateOfBirth
a1	Smith	John	42 Miller St	O'Connor	2602	A.C.T.	12-11-1970
a2	Neighan	Joanne	Brown Pl	Dickson	2604	ACT	8 Jan 1968
a3	Meyer	Marie	3/12-14 Hope Cnr	SYDNEY	2050	NSW	01-01-1921
a4	Smithers	Lyn	Browne St	DIXON	2012	N.S.W.	13/07/1970
a5	Nguyen	Ling	1 Milli Rd	Nrth Sydeny	2022	NSW	10/08/1968
a6	Faulkner	Christine	13 John St	Glebe	2037	NSW	02/23/1981
a7	Sandy	Robert	RMB 55/326 West St	Stuart Park	2713	NSW	7/10/1970

Database B

RecID	Name	Address	BYear	BMonth	BDay
b1	Meier, Mary	14 (App 3) Hope Corner, Sydney 2000	1927	4	29
b2	Janice Meyer	Bryan St, O'Connor ACT 2604	1968	11	20
b3	Jonny Smith	47 Miller Street, 2619 Canberra ACT	1970	12	11
b4	Lyng Nguyen	1 Millie Road, 2002 North Sydney, NSW	1968	8	10
b5	Kristina Fawkner	13 St John Street, 2031 Glebe	1981	2	23
b6	Bob Santi	55 East St; Stuart's Point; NSW 2113	1970	12	11
b7	Lynette Cain	6 / 12 Hope Corner, 2020 Sydney N.S.W.	1970	7	13

Fig. 2.2 Two small example database tables that are to be matched

matching have the same structure, and their content follows the same formats. It has been recognised that data cleaning and standardisation are crucial steps to successful data matching [78, 143]. The raw input data need to be converted into well-defined and consistent formats, and inconsistencies in the way information is represented and encoded need to be resolved [76, 224].

There are various factors that influence data quality, including different types of data entry errors, and the design of databases, such as the format and structure of their attributes. Some data quality factors are specific to personal information such as names and addresses. Name and address values are frequently entered either from handwritten forms using optical character recognition (OCR) software, read and typed, or typed as somebody speaks their personal information (possibly over the telephone). These different data entry modes can lead to typing, scanning, or phonetic errors [72]. How to deal with these challenges will be discussed in more detail in Chap. 3.

There are three (for certain types of data possibly four) major steps involved in data pre-preprocessing.

1. *Remove unwanted characters and words.* This step corresponds to an initial cleaning, where characters such as commas, colons, semicolons, periods, hashes, and quotes are removed. In certain applications, some words can also be removed if it is known that they do not contain any information that is of relevance to the data matching process. These words are also known as *stop words* [288].
2. *Expand abbreviations and correct misspellings.* This second step of data pre-processing is crucial to improve the quality of the data to be matched. Commonly

this step is based on look-up tables that contain name variations, nicknames, and common misspellings, and their correct or expanded versions. The standardisation of values conducted in this step will result in much reduced variations in attributes that contain name values.

3. *Segment attributes into well-defined and consistent output attributes.* This step deals with the common situation of database attributes that contain several pieces of information, such as the 'Address' attribute of the second database in Fig. 2.2. Finding a match between the content of this attribute and the content of the corresponding set of attributes in the first database ('Street', 'Suburb', 'Postcode' and 'State') is challenging. It is of advantage for data matching to split the content of attributes that contain several pieces of information into a set of new attributes that each contain one well-defined piece of information. The process of segmenting attribute values is also called *parsing* [143]. It is of high importance for both names and addresses, but also for dates. Various techniques have been developed to achieve such segmentation, either using rule-based systems or employing probabilistic techniques such as hidden Markov models [76]. These techniques will be covered in detail in Chap. 3.

4. *Verify the correctness of attribute values.* This last step can, for example, be employed for addresses if an external database is available that contains all known and valid addresses in a country or region. The detailed information in such an external database should include the range of street numbers, and the street name and type combinations that occur in towns and suburbs. Such a database will allow the verification of addresses and potentially even their correction, if for example it is known that there is no 'Miller Corner' in a certain town but only a 'Millers Court'. Applying such verification and correction might even be possible for name attributes, if, for example, a database of known residents is available that contains their full name and address details. However, because people can move, change their names, or might not even be registered (for example in a telephone directory), such name verification and correction might not help much to improve data quality. Rather, it might lead to wrong 'corrections' being introduced.

It is also possible, as illustrated in the pre-processed database tables in Fig. 2.3, to add attributes that are derived from existing attributes. For example, the gender of a person can often be correctly established from their given name (if a given name is distinctively used for males or females only). Similarly, if a postcode (or zipcode) value is missing in a record, its value could be extracted from the corresponding suburb or town name in case there is a unique postcode and suburb name combination.

It is important to note that the data pre-processing process must not overwrite the original input data. Once original values are overwritten (and if no backup has been made), then there is often no way to retrieve the original values in case a mistake was made during data pre-processing. Rather, new attributes should be created that contain the cleaned and standardised data. Ideally, data pre-processing is done in such a way that new database tables (or files) are generated that contain the cleaned and standardised data in such a format and structure that it can be easily used for the next step of the data matching process.

Database A – Cleaned and standardised

RecID	GivenName	Surname	Gender	StrPrefix	StrNum	StrName	StrType	Suburb	Postcode	State	BDay	BMonth	BYear
a1	john	smith	m		42	miller	street	oconnor	2602	act	12	11	1970
a2	joanne	neighan	f			brown	place	dickson	2604	act	8	1	1968
a3	mary	meier	f	3	12-14	hope	corner	sydney	2050	nsw	1	1	1921
a4	lynette	smithers	f			browne	street	dixon	2012	nsw	13	7	1970
a5	ling	nguyen	?		1	milli	road	north sydney	2022	nsw	10	8	1968
a6	christine	faulkner	f		13	john	street	glebe	2037	nsw	23	2	1981
a7	robert	sandy	m	rmb 55	326	west	street	stuart park	2713	nsw	7	10	1970

Database B – Cleaned and standardised

RecID	GivenName	Surname	Gender	StrPrefix	StrNum	StrName	StrType	Suburb	Postcode	State	BDay	BMonth	BYear
b1	mary	meier	f	apt 3	14	hope	corner	sydney	2000	nsw	29	4	1927
b2	janice	meier	f			bryan	street	oconnor	2604	act	20	11	1968
b3	john	smith	m		47	miller	street	canberra	2619	act	11	12	1970
b4	lyng	nguyen	?		1	millie	road	north sydney	2002	nsw	10	8	1968
b5	kristina	fawkner	f		13	saint john	street	glebe	2037	nsw	23	2	1981
b6	robert	santi	m		55	east	street	stuarts point	2113	nsw	11	12	1970
b7	lynette	cain	f	6	12	hope	corner	sydney	2020	nsw	13	7	1970

Fig. 2.3 The pre-processed (cleaned and standardised) versions of the two database tables from Fig. 2.2. Both databases now consist of the same attributes. The format and content of these attributes have been standardised in that various punctuations were removed, all letters were converted into lower case, nicknames replaced by the corresponding proper names, typographical errors corrected, dates and addresses were split into several well-defined fields, and contradicting data corrected (such as the postcode for suburb 'Glebe' which has a correct value of '2037' and not '2031', as was recorded in the original record 'b5'). Additionally, the attribute 'Gender' was added. Its values are based on the given name values of the corresponding records only for given names that are known to be distinctively male or female

2.3 Indexing

The cleaned and standardised database tables (or files) are now ready to be matched. Potentially, each record from one database needs to be compared with all records in the other database to allow the calculation of the detailed similarities between two records. This leads to a total number of record pair comparisons that is quadratic in the size of the databases to be matched. Matching the example databases from Fig. 2.3 leads to a total of $7 \times 7 = 49$ comparisons (between one record from database **A** and one record from database **B**).

Clearly, this naïve comparison of all record pairs does not scale to very large databases. Matching two databases with one million records each (as are common in many public and private sector organisations today) will result in $1,000,000 \times 1,000,000 = 1,000,000,000,000$, i.e. one trillion, record pair comparisons. Even if $100,000$ comparisons can be performed in one second ($10\,\mu s$ or 0.01 ms per comparison), it would take $2,777.78$ h, or nearly 116 days, to compare these two databases.

The majority of the comparisons will be between two records that are clearly not matches. As can be seen from Fig. 2.3, most record pairs have no or only a small number of attribute values that are equal or highly similar with each other. For example, record 'a1' in database **A** has the same year of birth (1970) as records 'b3', 'b6', and 'b7' from database **B**, but it only has two other attribute values in common with record 'b6' (gender 'm' and street type 'street'), and no other attribute value in common with record 'b7'.

It is generally the case that when matching two databases, the potential number of comparisons grows quadratically with the number of records in the databases to be matched, while the number of possible true matches only increases linearly. This is because it is likely that one record from database **A** only matches to a small number of records from database **B**. In the case where both databases **A** and **B** do not contain duplicate records (i.e. several records that refer to the same entity), then the maximum number of true matches that are possible is always smaller than or equal to the number of records in the smaller of the two databases.

To reduce the possibly very large number of pairs of records that need to be compared, *indexing* techniques are commonly applied [64]. These techniques filter out record pairs that are very unlikely to correspond to matches. They generate *candidate record pairs* that will be compared in more detail in the comparison step of the data matching process to calculate the detailed similarities between two records, as will be described in the following section.

Various indexing techniques for data matching and deduplication have been developed [64]. The traditional approach to indexing is called *blocking* [20]. It splits each database into smaller blocks according to some *blocking criteria* (generally known as a *blocking key*). Only records from the two databases that have been inserted into the same block, i.e. who share the same value for a blocking criteria (have the same blocking key value), are compared with each other. An example blocking criteria could be that records that have the same postcode value are inserted into the same block, while another blocking criteria could be that records that have the same phonetically encoded surname value are inserted into the same block. Such phonetic encoding algorithms, like for example *Soundex* [57], are commonly used in the indexing step to ensure that records are inserted into the same blocks even if they have some typographical variations in the value of their blocking criteria. Chapter 4 will discuss traditional blocking and several other indexing techniques in more detail, and also provide an experimental evaluation of these techniques to illustrate their performance on different types of data.

When the traditional blocking technique is applied to the cleaned and standardised databases from Fig. 2.3 using the two blocking criteria (1) Soundex of surname values ('Sndx-SN') and (2) taking the first three digits of postcode values ('F3D-PC'), then the blocks and candidate record pairs shown in Figs. 2.4 and 2.5 are generated. As can be seen, from the full number of 49 record pairs (without indexing), only 12 candidate record pairs are generated. These candidate pairs will be compared in detail, as will be described in the following section.

Looking at the candidate record pairs generated and comparing the corresponding record pairs in Fig. 2.3, one can see that this specific blocking approach selects most of the record pairs that likely refer to a match, such as (a1, b3) ('John Smith'), (a3, b1) ('Mary Meier'), and (a5, b4) ('Ling Nguyen'). This blocking approach does, however, miss the pair (a4, b7) ('Lynette Smithers' / 'Lynette Cain') which is possibly

Database A – Blocking information

RecID	Surname	Sndx-SN	Postcode	F3D-PC
a1	smith	s530	2602	260
a2	neighan	n250	2604	260
a3	meier	m600	2050	205
a4	smithers	s536	2012	201
a5	nguyen	n250	2022	202
a6	faulkner	f425	2037	203
a7	sandy	s530	2713	271

Database B – Blocking information

RecID	Surname	Sndx-SN	Postcode	F3D-PC
b1	meier	m600	2000	200
b2	meier	m600	2604	260
b3	smith	s530	2619	261
b4	nguyen	n250	2002	200
b5	fawkner	f256	2037	203
b6	santi	s530	2113	211
b7	cain	c500	2020	202

Fig. 2.4 The blocking key values (BKVs) generated from the two database attributes 'Surname' and 'Postcode'. For surnames, BKVs are generated by applying Soundex encoding [57] on surname values (labelled 'Sndx-SN'), while the BKVs of postcodes are generated by taking their first three digits only (labelled 'F3D-PC')

Candidate record pairs generated from Surname blocking

BKVs	Candidate record pairs
m600	(a3, b1), (a3, b2)
n250	(a2, b4), (a5, b4)
s530	(a1, b3), (a1, b6), (a7, b3), (a7, b6)

Candidate record pairs generated from Postcode blocking

BKVs	Candidate record pairs
202	(a5, b7)
203	(a6, b5)
260	(a1, b2), (a2, b2)

(a1, b2)
(a1, b3)
(a1, b6)
(a2, b2)
(a2, b4)
(a3, b1)
(a3, b2)
(a5, b4)
(a5, b7)
(a6, b5)
(a7, b3)
(a7, b6)

Fig. 2.5 The candidate record pairs generated from the BKVs that occur in both database **A** and **B**. The table on the right-hand side shows the union of all generated candidate record pairs

the same woman because both records have the same given name and the same date of birth. This woman might have married and changed her surname and address, and is therefore missed by the two blocking criteria used. This example highlights the careful need for domain and data matching knowledge when defining blocking criteria. Both the quality and completeness, as well as the frequency distribution of the values in an attribute need to be considered when attributes are selected to be used as blocking keys. These issues will be further discussed in Chap. 4.

2.4 Record Pair Comparison

The candidate record pairs that were generated in the indexing step require detailed comparisons to determine their overall similarity. Generally, the similarity between two records is calculated by comparing several record attributes. Ideally, not just the attributes used in the indexing step are used for this, but also other attributes that are available in the databases that are matched. In the running example used in this chapter, while the blocking was based on the 'Surname' and 'Postcode' attributes, the comparison should for example also include the attributes 'GivenName', 'Street-Num', 'StreetName', 'Suburb', and the three date of birth attributes. The more similar values two records have in common across these attributes, the more likely it will be that they correspond to the same individual.

Even after records have been cleaned and standardised, it is possible that there are different attribute values in the records that correspond to true matches (i.e. that refer to the same entity). In the example, the records 'a6' and 'b5' very likely correspond to the same individual. However, the given name, surname and street name values of these two records are all slightly different. Rather than only conducting exact matching between attribute values, it is therefore essential to conduct some form of approximate comparison that for a compared pair of attribute values returns a measure of their similarity.

Generally, similarity values are normalised numerical values, with a similarity of 1.0 corresponding to an exact match between two attribute values, a similarity of 0.0 corresponding to a total dissimilarity between two values, and similarities in-between 0.0 and 1.0 corresponding to some degree of similarity between two attribute values. Figure 2.6 shows the similarities calculated between attribute values for the 12 candidate record pairs from Fig. 2.5.

Given different attributes contain various types of data, different approximate similarity comparison functions are required [61]. For attributes that contain string values, such as names and addresses, a large number of approximate string comparison functions is available [57]. Specific comparison functions for dates, ages, times, locations and numerical values are used for attributes that contain such data [61]. For certain sets of attributes, such as given names, surnames, or dates (consisting of a day, month and year value), it is also advisable to compare attributes as a group rather than only individually. For example, for names from several Asian cultures, certain name values can interchangeably be used as given name and surname (such as 'Qing Yang' and 'Yang Qing'). Therefore, comparing the given name value from one record with the surname value from another record, and the other way round, will help to detect pairs of records where these two name components have been swapped. Similarly, dates can have their day and month values swapped as they are recorded either following the American date format (MM/DD/YYYY) or the format used in many other countries (DD/MM/YYYY). Chapter 5 covers a large number of different comparison functions for different types of data, and highlights various issues that need to be considered when using for example names and addresses for data matching.

For each candidate record pair several attributes are generally compared, resulting in a vector of numerical similarity values for each pair. These vectors are called *comparison vectors*. They will be used in the classification step to decide if a record pair is classified as a match or a non-match.

The comparison vectors resulting from the comparison of the 12 candidate records pairs of the running example are shown in Fig. 2.6. Different approximate comparison functions were used. The sum of all similarity values for each comparison vector is

RecID	GivenName	Surname	StrNum	StrName	Suburb	BDay	BMonth	BYear	SimSum
a1	john	smith	42	miller	oconnor	12	11	1970	
b2	janice	meier		bryan	oconnor	20	11	1968	
	0.61	0.6	0.0	0.0	1.0	0.0	1.0	0.5	3.71
a1	john	smith	42	miller	oconnor	12	11	1970	
b3	john	smith	47	miller	canberra	11	12	1970	
	1.0	1.0	0.5	1.0	0.6	0.5	0.5	1.0	6.10
a1	john	smith	42	miller	oconnor	12	11	1970	
b6	robert	santi	55	east	stuarts point	11	12	1970	
	0.47	0.6	0.0	0.0	0.31	0.5	0.5	1.0	3.39
a2	joanne	neighan		brown	dickson	8	1	1968	
b2	janice	meier		bryan	oconnor	20	11	1968	
	0.78	0.56	0.0	0.73	0.51	0.0	0.5	1.0	4.08
a2	joanne	neighan		brown	dickson	8	1	1968	
b4	lyng	nguyen	1	millie	north sydney	10	8	1968	
	0.47	0.64	0.0	0.0	0.45	0.0	0.0	1.0	2.56
a3	mary	meier	12-14	hope	sydney	1	1	1921	
b1	mary	meier	14	hope	sydney	29	4	1927	
	1.0	1.0	0.4	1.0	1.0	0.0	0.0	0.75	5.15
a3	mary	meier	12-14	hope	sydney	1	1	1921	
b2	janice	meier		bryan	oconnor	20	11	1968	
	0.47	1.0	0.0	0.0	0.44	0.0	0.5	0.5	2.91
a5	ling	nguyen	1	milli	north sydney	10	8	1968	
b4	lyng	nguyen	1	millie	north sydney	10	8	1968	
	0.83	1.0	1.0	0.94	1.0	1.0	1.0	1.0	7.78
a5	ling	nguyen	1	milli	north sydney	10	8	1968	
b7	lynette	cain	12	hope	sydney	13	7	1970	
	0.6	0.47	0.5	0.0	0.5	0.5	0.0	0.5	3.07
a6	christine	faulkner	13	john	glebe	23	2	1981	
b5	kristina	fawkner	13	saint john	glebe	23	2	1981	
	0.81	0.87	1.0	0.45	1.0	1.0	1.0	1.0	7.12
a7	robert	sandy	326	west	stuart park	7	10	1970	
b3	john	smith	47	miller	canberra	11	12	1970	
	0.47	0.47	0.0	0.0	0.54	0.0	0.5	1.0	2.98
a7	robert	sandy	326	west	stuart park	7	10	1970	
b6	robert	santi	55	east	stuarts point	11	12	1970	
	1.0	0.73	0.0	0.83	0.78	0.0	0.5	1.0	4.85

Fig. 2.6 Similarity values (comparison vectors) calculated using different approximate similarity comparison functions for the 12 candidate record pairs from Fig. 2.5. For attributes containing names, the Jaro–Winkler [215] approximate string comparison function was used, while for the attributes that contain numbers the edit distance [89] function was employed

shown on the right-hand end of each vector (SimSum). These sums can be used for a simple threshold-based classification approach as will be described in the following section.

2.5 Record Pair Classification

Classifying the compared record pairs based on their comparison vectors or their summed similarities is a two-class (binary) or three-class classification task. In the two-class case, each compared record pair is classified to be either a *match* or a *non-match*. The first class contains the pairs of records that are assumed to refer to the same real-world entity, while for the second class it is assumed that the two records in a pair do not refer to the same entity. All record pairs that were removed by the indexing step and that were not compared in the comparison step are implicitly classified as non-matches.

In traditional data matching approaches, for example those based on probabilistic record linkage [108, 143], record pairs are classified into one of three classes, rather than only matches and non-matches. The third class are the *potential matches*. These are the record pairs where the classification outcome is not clear, and where a manual *clerical review* [143] is required to decide the final match status.

Most research in data matching in the past decade has concentrated on improving the classification accuracy of record pairs. Various machine learning techniques have been investigated, both unsupervised and supervised [31, 59, 85, 102]. So called *active learning* techniques have also been investigated [231, 252]. With these techniques, a subset of (difficult to classify) record pairs is given for manual assessment and classification (into matches and non-matches), and the resulting classified record pairs are used to re-train a new and improved classifier. After several iterations, this process can achieve an improved matching accuracy with much reduced manual efforts compared to the traditional approach of full manual clerical review of all potential matches.

The classification of each compared record pair can be based on either the full comparison vectors or on only the summed similarities. Figure 2.6 shows the 12 compared record pairs, their comparison vectors and their summed similarities (SimSum). The maximum possible summed similarity (of two records that are matching exactly on all compared attributes) would be 8.0, because eight attributes are compared each returning a similarity between 0 and 1. Figure 2.7 shows the outcomes of a simple threshold-based classifier where all compared record pairs with a SimSum value equal to or above 6 are classified as matches, all pairs with a SimSum value between 4 and 6 as potential matches, and all other pairs as non-matches. As a result, the three pairs (a1, b3), (a5, b4) and (a6, b5) will (presumably) be correctly classified as matches. Of the three potential match pairs (a2, b2), (a3, b1) and (a7, b6) given for manual clerical review, the second pair (a3, b1) will likely be classified as a match, while the other two pairs might be classified as non-matches. Figure 2.8 shows the actual records of the three pairs that were classified as matches.

Candidate pair	SimSum	Classification
(a1, b2)	3.71	Non-match
(a1, b3)	6.10	Match
(a1, b6)	3.39	Non-match
(a2, b2)	4.08	Potential match
(a2, b4)	2.56	Non-match
(a3, b1)	5.15	Potential match
(a3, b2)	2.91	Non-match
(a5, b4)	7.78	Match
(a5, b7)	3.07	Non-match
(a6, b5)	7.12	Match
(a7, b3)	2.98	Non-match
(a7, b6)	4.85	Potential match

Fig. 2.7 Three-class classification of the compared record pairs from Fig. 2.6 into matches (Sim-Sum $\geq$ 6.0), non-matches (SimSum $\leq$ 4.0) and potential matches (6.0 > SimSum > 4.0)

Database A

RecID	Surname	GivenName	Street	Suburb	Postcode	State	DateOfBirth
a1	Smith	John	42 Miller St	O'Connor	2602	A.C.T.	12-11-1970

Database B

RecID	Name	Address		BYear	BMonth	BDay
b3	Jonny Smith	47 Miller Street, 2619 Canberra ACT		1970	12	11

Database A

RecID	Surname	GivenName	Street	Suburb	Postcode	State	DateOfBirth
a5	Nguyen	Ling	1 Milli Rd	Nrth Sydeny	2022	NSW	10/08/1968

Database B

RecID	Name	Address		BYear	BMonth	BDay
b4	Lyng Nguyen	1 Millie Road, 2002 North Sydney, NSW		1968	8	10

Database A

RecID	Surname	GivenName	Street	Suburb	Postcode	State	DateOfBirth
a6	Faulkner	Christine	13 John St	Glebe	2037	NSW	02/23/1981

Database B

RecID	Name	Address		BYear	BMonth	BDay
b5	Kristina Fawkner	13 St John Street, 2031 Glebe		1981	2	23

Fig. 2.8 The three record pairs that were classified as matches

The traditional approaches to record pair classification have the problem that each record pair is classified independently of all others pairs based only on its comparison vector (or its summed similarity). As a result, a single record from one database can be matched with several records from the other database. In certain applications this might not be permitted, for example if it is known that the two databases that are matched each only contain one record per entity (i.e. no duplicate records). Recent research into *collective* classification techniques for data matching has aimed to overcome this drawback by classifying record pairs not only based on their pair-wise similarities, but also using information on how records are related or linked to other records. These approaches apply relational clustering or graph-based techniques [31, 155, 272] to generate a global decision model. Much improved matching results have been achieved with these collective classification techniques. Their computational complexities, however, make scaling these techniques to the matching of very large databases challenging [142]. These techniques, as well as the more traditional pair-wise classification techniques, will be presented in detail in Chap. 6.

2.6 Evaluation of Matching Quality and Complexity

Once the compared record pairs are classified into matches and non-matches, the quality of the identified matches needs to be assessed. Matching quality refers to how many of the classified matches correspond to true real-world entities, while matching completeness is concerned with how many of the real-world entities that appear in both databases were correctly matched [71]. As will be discussed in detail in Chap. 7, accuracy measures such as precision and recall, that are also used in fields such as data mining, machine learning, and information retrieval, are commonly used to assess matching quality

Both matching accuracy and completeness are affected by all steps of the data matching process, with data pre-processing helping to make values that are different to each other more similar, indexing filtering out pairs that likely are not matches, and the detailed comparison of attribute values providing evidence of the similarity between two records. While the accuracy of data matching is mostly influenced by the comparison and classification steps, the indexing step will impact on the completeness of a data matching exercise because record pairs filtered out in the indexing step will be classified as non-matches without being compared.

The complexity of a data matching or deduplication project is generally measured as the number of candidate record pairs that are generated by an indexing technique compared to the number of all possible pairs that would be generated in the naive. matching where no indexing is applied. For the running example shown in this chapter, the naive full pair-wise comparison of all records from database **A** with all records from database **B** would result in $7 \times 7 = 49$ record pair comparisons. The indexing (blocking) applied in this example has reduced this number to the 12 candidate pairs shown in Fig. 2.5. This corresponds to a reduction of over 75 %.

To evaluate the completeness and accuracy of a data matching project, some form of ground-truth data, also known as *gold standard*, are required. Such ground-truth data must contain the true match status of all known matches (the true non-matches can be inferred from them). However, obtaining such ground-truth data is difficult in many application areas. For example, when matching a large tax payers database with a social security database it is usually not known which record pair classified as a match refers to a real, existing individual who has a record in both databases.

Only further investigations, such as checking extra data about the individual under consideration, or even contacting them, can help determine the truth about such a classified match.

A related problematic issue is the manual classification of potential matches through clerical review. It is often difficult to make a manual match or non-match decision with high confidence if the two records in a potential match pair contain several attribute values that differ from each other. Without further external information, a decision that was made manually might be wrong. Additionally, the manual classification of a large number of potential matches is a time-consuming, cumbersome and error-prone process. Assuming that the manually classified potential matches can be used as training data for supervised classification or even as gold standard for another data matching project is dangerous. The issues relevant to evaluating data matching will be discussed in detail in Chap. 7.

As the classification results in Fig. 2.7 show, even the matching of two small example databases results in a quite imbalanced distribution of matches to non-matches (four matches to eight non-matches after clerical review in this example). This class-imbalance gets much worse as larger databases are being matched. The number of matches generally grows linear (or even sublinear), while the number of non-matches (even after indexing) grows subquadratic [71], as will be discussed further in Chap. 6. When evaluating the results of a data matching or deduplication project, even when ground-truth data are available, care must be taken. The normal accuracy measure that is generally used for many classification tasks is not recommended. Various measures that are suitable for assessing the quality and complexity of data matching and deduplication will be presented in detail in Chap. 7.

2.7 Further Reading

The data matching process (with some variations to the steps described in this chapter) is discussed in most books that cover data matching [19, 143, 195, 249], as well as in several reports and overview articles [103, 119]. A recent survey of indexing techniques is provided by Christen [64], while many surveys have been written over the past decades on approximate string comparisons techniques [57, 84, 133, 152, 175, 196]. On the other hand, while many different classification techniques have been explored within the domain of data matching, only a few publications have comparatively evaluated several techniques [59, 102, 168]. The issues involved in evaluating data matching results are being discussed in two recent publications [71, 187].

Part II
Steps of the Data Matching Process

Chapter 3
Data Pre-Processing

3.1 Data Quality Issues Relevant to Data Matching

Most real-world databases contain noisy, inconsistent and missing data due to a variety of factors [19, 135, 218]. It is generally accepted that low data quality costs businesses and governments billions of lost revenue every year. It has been estimated that data quality problems can result in up to 12 % lost revenue for businesses [177]. For any type of data analysis, processing and management, the *garbage-in garbage-out* principle holds. If the quality of the input data is low, then the output generated is normally not of high quality or accuracy either.

A large body of work has covered the various issues involved with data quality in depth [19, 177, 218]. There are several dimensions to data quality. The ones relevant to data matching are:

- *Accuracy.* How accurate are the attribute values in the database(s) used for matching or deduplication? Is it known how the data have been entered or recorded? Have data entry checks been performed, and have the data been verified for correctness using external reference data (such as references of known and valid addresses)?
- *Completeness.* How complete are the data? How many attribute values are missing in the databases used? Is it known why certain attribute values are missing? Are attributes missing that would be of use for data matching?
- *Consistency.* How consistent are the values within a single database used for matching or deduplication, and how consistent are values across two or more databases used for matching? The format and coding of individual attributes even within a single database can change over time. Is it known if the databases contain duplicate records for the same entity (for example because a person moved to a different address and therefore was recorded as a new separate customer)?
- *Timeliness.* How old are the data available? For the matching of two databases, have the data been recorded at the same time or not? This can be a crucial factor to a successful matching because personal information, such as people's addresses, telephone numbers, and even names, change over time. If the data to be matched

P. Christen, *Data Matching*, Data-Centric Systems and Applications,
DOI: 10.1007/978-3-642-31164-2_3, © Springer-Verlag Berlin Heidelberg 2012

have been recorded at different points in time then this needs to be taken into account during the data matching process.

- *Accessibility*. Are all the data required available in the database to be deduplicated or the databases to matched? Is there enough information in the form of attributes that cover different aspects of the entities in the databases to allow detailed comparisons and accurate classification? If for example only names but no address information is available then accurate matching of two large databases will be impossible because many records might contain the names 'John Smith' or 'Mary Miller'.
- *Believability*. Can the values stored in the databases be regarded as credible or true? Or is it possible that values are wrong or impossible?

Arguably the most important data quality dimensions for data matching and deduplication are *accuracy* and *consistency*, because a large portion of efforts in the indexing, comparisons and classification steps (that will be covered in Chaps. 4–6) deal with inaccurate and inconsistent data. If data would be of perfect quality, then data matching could be accomplished through straightforward database join operations and no sophisticated indexing techniques or approximate comparison functions would be needed. As long as data are of imperfect quality, however, techniques are needed that can deal with inaccurate and inconsistent data while still achieving high matching quality.

Various root causes for data quality problems have been identified [177]. The ones that are relevant to data matching are:

- *Multiple data sources*. If data are recorded by different organisations or different systems, at different locations, at different points in time, or using different data entry modes [72], then such data will likely be inconsistent.
- *Subjective judgement of data production*. If certain aspects of the entities in the databases to be matched were not recorded because they were deemed not to be of importance, then this information will be missing. This can potentially hamper data matching, if not enough data are available to accurately compare and classify pairs or groups of records. For example, if dates of birth have not been recorded then the matching of two hospital patient databases might have to rely upon patient's name and address details only.
- *Limited computing resources*. As will be discussed in Chap. 4, data matching is a computationally expensive process. If the databases to be matched are large and not enough computing and storage power are available, then it might not be feasible to run a sophisticated and accurate data matching algorithm. The results achieved with a simpler matching algorithm might not be accurate enough for certain applications. The use of cloud computing resources might be difficult for data matching because of privacy and confidentiality concerns.
- *Security/accessibility trade-off*. This root condition is highly relevant when databases that contain personal information are to be matched across organisations. Privacy regulations or security concerns might prevent that data which contain personal information can be accessed, thereby preventing certain data matching

projects. The topic of privacy within the context of data matching will be covered in detail in Chap. 8.

- *Coded data across disciplines.* This condition will affect the consistency of data between different databases. If the databases to be matched originate in different organisations or different disciplines, then careful mapping between different formats and encodings is required before any matching can be attempted.

- *Complex data representations.* Many traditional data matching algorithms can only be applied on data that are made of strings (such as name and address values) or numerical values (such as dates or age values). Increasingly, however, entity information is stored using more complex representations, such as XML schemas [270], or it consists of different types of entities that are potentially linked with other entities. Multi-relational, normalised databases are commonly used to represent different types of entities in an organisation and their interactions or relationships. Data matching algorithms must be able to deal with such types of complex data. This topic will be further covered in Sect. 9.2.

- *Volume of data.* As the size of the databases held by many organisations are ever increasing, deduplicating or matching them becomes more challenging, because more computing resources and more time is required. Chapter 4 deals with the topic of indexing for data matching, which is aimed at making the data matching process more scalable to very large databases.

- *Input rules too restrictive or bypassed.* This root cause can result in data of low quality because data are entered into fields or attributes that originally had a different purpose. For example, assume an emergency department's patient database where the personal details of emergency patients are recorded. The design of the system does require a valid date of birth to be entered for each patient's record. Imagine some patients arriving semiconscious or unconscious and without any identification documents. In such cases, no detailed information about their date of birth will be available. The receptionist or nurse who enters the data will likely be under time pressure. A simple solution for them to enter a valid record is to guess a patient's age and to then enter a date of birth value with the day and month values set to '01'. As a result, an unexpected high percentage of records in such a database will have a date of birth of 1st January.

- *Changing data needs.* The information need of organisations often changes over time, as they adapt to new regulations, implement new information systems, restructure themselves, or as they merge with other organisations. Only data that are useful and relevant for the operation of an organisation are normally collected, and therefore what information is stored in databases changes over time. New fields or attributes might be added to a database, attributes no longer considered relevant might be removed, or formats and codes might change over time. If data that have been recorded over time (or at different points in time) are being matched or deduplicated, then these changes can make the matching challenging, because only the information in attributes commonly available across time can be employed in the matching process.

- *Distributed heterogeneous systems.* Data recorded and stored in different systems potentially have different formats, different types and different values. When such

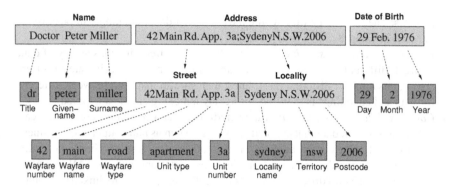

Fig. 3.1 An example of data pre-processing applied to one record consisting of personal details. Cleaning includes removing unwanted characters and converting all letters into lower case. Standardisation consists of correcting typographical errors such as replacing 'sydeny' with 'sydney', and replacing abbreviations with standard forms. The third step is the segmentation of the input into well-defined output fields that are then used as the actual attribute values in the deduplication or data matching process

data are being matched, a careful analysis prior to matching is required to make sure that the same type of information (that will be compared in detail between records) is available in the same format and structure.

The remainder of this chapter covers in more detail how values in the input database(s) can be pre-processed to make them suitable for data matching and deduplication, as illustrated in Fig. 3.1. Data pre-processing for data matching consists of four major steps, as will be discussed in Sect. 3.5. First, however, discussions on the specific characteristics that names and other personal information pose to data quality, and where variations and errors in names come from, are needed.

3.2 Issues with Names and Other Personal Information

Names and other personal details play a crucial role in daily life because people are using them to identify individuals, ranging from family and friends, to work colleagues, and all the way to politicians and celebrities. For organisations both in the private and public sectors, names are often a primary source of identification of the individuals they are in contact with.

Personal names are a major component of the information used in many data matching or deduplication processes to identify records that refer to the same individuals. A large amount of the data collected by businesses and governments are about people. The identifying data collected about individuals generally include their names and addresses, dates of birth, social security or drivers license numbers, telephone numbers, and email addresses.

Much of the daily news fed to us through different channels is also about people, and therefore names commonly appear in news articles, on Web sites, and even most scientific and technical documents include their authors' names, affiliations, and other contact details.

Personal names are frequently used in Web searches to find information about individuals, in online stores to find movies, songs, albums or books by certain artists or writers, and when querying digital libraries to find articles or documents written by a specific author. The ten most popular query terms used with the Google Web search engine over the past decade include several personal names (of certain popular celebrities).[1]

Personal names and other identifying details have characteristics that make them different from general text [40, 208, 210]. These characteristics need to be considered when databases are matched or deduplicated, because they will influence how efficient and accurate the matching can be conducted. The following list highlights some of the issues with names, with an emphasis on the characteristics of names from English speaking and other Western countries.

- While in many languages for general words there is only one correct form, there are often several variations for what is seen as the same personal name. For example, there are more than forty variations of 'Amelia'[2]: 'Aemelia', 'Aimiliona', 'Amalea', 'Amalee', 'Amaleta', 'Amalia', 'Amalie', 'Amalija', 'Amalina', 'Amaline', 'Amalita', 'Amaliya', 'Amaly', 'Amalya', 'Amalyna', 'Amalyne', 'Amalyta', 'Amelie', 'Amelina', 'Ameline', 'Amelita', 'Ameliya', 'Amelya', 'Amelyna', 'Amelyne', 'Amelyt', 'Amilia', 'Amy', 'Delia', 'Em', 'Emelie', 'Emelina', 'Emeline', 'Emelita', 'Emi', 'Emma', 'Emmeline', 'Emmi', 'Emmie', 'Emmy', 'Mali', 'Malia', 'Malika', 'Meelia', 'Melia', 'Meline', 'Millie' and 'Milly'.

- In daily life, people often use or are given nicknames, rather than the name they were given by their parents at birth. Such nicknames can be short forms of their given name (such as 'Liz' for 'Elizabeth', 'Tina' for 'Christina', or 'Bob' for 'Robert'), they can be a variation of their surname of family name (such as 'Vesty' for 'Vest'), or their nickname is based on some life event, physical characteristic ('Ginger' for a red-haired person), or a character sketch of an individual [40]. Matching such nicknames can obviously be much more difficult than matching small name variations like the ones shown above. In certain cases, it will be impossible to find a match on a nickname at all.

- There are generally no legal regulations of what constitutes a name, with only some specific restrictions with regard to religious, political, or historical characters in certain countries.

- Names are language and culture specific [208]. In Anglo-Saxon countries (including the UK, USA, Canada, South Africa, Australia, Ireland, and New Zealand), names are made of a given or first name and a surname or family name, with an optional middle name (or initial) in-between, and possibly a name prefix or suffix

---

[1] See: http://www.google.com/press/zeitgeist.html.

[2] See: http://www.thinkbabynames.com/meaning/0/Amelia.

(such as 'Jr' or 'Snr'). In several European countries compound names are common, such as 'Hans-Peter' in Germany or 'Jean-Pierre' in France. Hispanic names often consist of two surnames.

- People can change their names over time, most commonly when they get married or divorced. While traditionally in many western countries a wife will take on the surname of her husband, this tradition is changing rapidly and today a husband might decide to take on his wife's surname. Alternatively, a couple might decide to compound their two surnames. For example, when 'Sally Smith' marries 'John Miller' she changes her name to 'Sally Smith-Miller', while her husband changes his name to 'John Miller-Smith'. If they have children, they need to decide which compound surname to give to their children.
- Outside of English speaking or Western cultures, each language has its own names and its own naming conventions, with cultures within the same language having their own ways of how names are selected for babies when they are born, and how they can change over an individual's life-time [40, 208].
- For languages that are based on characters different to the Roman alphabet, the way names are transliterated into the Roman alphabet is crucial. There might be several standards for transliterating for example Chinese, Japanese, Korean, Thai or Arabic names into the Roman alphabet, leading to variations of the same name. Individuals who are unfamiliar with standard transliteration systems might decide on a Roman version of their name in an ad hoc fashion, or alternatively choose or add a Western given name to their full name to better fit into a Western culture [208]. Arabic names commonly consist of several components and can contain various prefixes and suffixes that can be separated by hyphens or whitespaces, and that change over an individuals life-time depending upon his or her circumstances.

All these issues make data matching or deduplication using personal names a challenging undertaking, because the name values for the same individual might differ across two databases, or even within a single database. In our increasingly multicultural world where people are more mobile than ever before, where international travels and living in a country different to one's home country are common, and with the globalisation of businesses, the appropriate cleaning and standardisation of names in databases used for data matching are crucial components to achieve accurate matching results.

Besides names, addresses of where people live or where businesses are located, are a second major component of the information used in data matching [76]. While addresses are generally more standardised than names, there are several specific issues that need to be considered.

Addresses in most countries consists of a locality component and a street component, as illustrated in Fig. 3.1. The locality component generally contains a postcode or zipcode which allows mail to be efficiently directed to the destination locality. Postcodes and zipcodes are determined by a country's postal organisation. In some countries, such as Australia, each postcode covers an area of roughly the same number of households or businesses in order to allow a balanced handing of postal mail. However, as new suburbs are being built and existing areas change their characters, postcode boundaries do change over time, and new postcodes are being generated. In other countries, the area of individual postcodes can be vastly different from others, and postcode boundaries do not change even when populations change.

The street component of an address usually consists of a street number, street name, and a street type. Additional street address elements can include flat or apartment numbers, floor numbers, and business or institution names. Alternatives to street addresses are post boxes and road-side mailboxes. While postal services in individual countries generally provide guidelines or standards of how a mailing address should be written, even if an address on a letter or parcel does not follow such guidelines, the item generally still arrives at its destination because the post man or woman or courier uses their local knowledge when delivering mail.

Because people know their mail arrives even if the address they provide is not totally accurate, a phenomena that has been reported is that individuals who reside close to an area that has a higher social status (for example if it is known that more rich people or celebrities live in that area) commonly use the name of the more prestigious area rather than the name of the area they live in, in order to impress friends and family. It is unlikely however that they would use such inaccurate address details when providing information to government agencies.

The third component of personal information that is commonly used for data matching are dates, such as dates of birth, dates of death, travel dates, or dates of admission to a hospital, to name a few. The major issue with dates is that if an individual does not know or does not remember a date when required, then commonly some approximation of the true date is being recorded. This might happen when dates are required from elderly people, or individuals need to report dates of other family members. If an accurate date is unknown, a common placeholder is to use the first day of the month (if the month of an event is known), or even the first day of January if only the year of when an event occurred is known.

Both personal names and people's addresses will likely change over time. Today, though unlikely, even the gender of a person can change. The only pieces of demographic information for an individual that do not change are their date and place of birth (that is why these two pieces of information are recorded on passports).

3.3 Types and Sources of Variations and Errors in Names

Given the many issues on name variations covered in the previous section, some discussion about studies that have investigated names variations and errors is required.

In an early study on spelling errors in general words, Damerau found that the majority of errors, over 80 %, were single character errors [89]. These were either a single letter that was deleted, an extra letter that was inserted, a letter that was substituted with another letter, or two adjacent letters that were transposed. The most common type of errors were character substitutions, followed by character deletions, then character insertions and finally the transposition of two characters. Multiple errors in a word were even less frequent than character transpositions. Damerau's work lead to the development of edit distance based approximate string comparison function that aim to overcome such character-based variations, as will be described in Sect. 5.3.

Several other studies that followed from Damerau's work have reported similar results with regard to the types and distributions of variations or errors [133, 172, 214]. However, a more recent study that investigated patient names within hospital databases found different types and distributions of variations [113]. The most common type of variation, with 36 %, in these data were the insertion of an additional name word, initial or title word. The second most common type with 14 % were differences of several characters due to spelling mistakes or the use of nicknames. Other types of variation were differences in punctuation (like in 'O'Brian', 'OBrian' or 'O Brian') with 12 % of all variations, and changed surnames for female patients with 8 % of all variations. In this particular study, single character variations only accounted for 39 % of all variations compared to the over 80 % reported by Damerau [89]. This study highlights the differences between names and general text that was discussed in the previous section.

These differences need to be considered when data matching algorithms are being developed and employed on data that contain personal names. The most commonly occurring variations and errors can be categorised into [175, 243]:

- Spelling variations due to typographical errors that do not affect the phonetical structure of a name, such as 'Meier' and 'Meyer', or 'Christina' and 'Kristina'. These variations still pose a problem for data matching and need to be dealt with.
- Phonetic variations where the phonemes are modified for example through mishearing during data entry, and the structure of a name is changed substantially, such as from 'Sinclair' to 'St. Clair'.
- Double names that might be given in full, only the first name but not the middle name, or given as compound names (like 'Peter Paul Miller', 'Peter Miller', 'Peter Paul-Miller' or 'Peter-Paul Miller'. The variations here include potential different separators, missing name components, or even swapped name components.
- Name alternatives such as nicknames, married names or other deliberate name changes; and initials only (mainly for given and middle names).

A survey on spelling correction by Kukich has provided further details about character level misspellings that occur during data entry of general text [172]. She described three types of errors: (1) typographical errors, where the assumption is that the individual who was doing the data entry knew the correct spelling of a word but made a typing mistake (this author's favourite such mistake is to type 'Sydeny' instead of 'Sydney'); (2) cognitive errors, which are assumed to come from a lack of knowledge of the correct spelling or from misconceptions; and (3) phonetic errors, coming from the substitution of a correct spelling with a similar sounding one that is also correct.

The second and third type of errors will be a major cause for name variations, such as the many variations of the name 'Amelia' on p. 43, in databases where values are entered manually. The combination of spelling variations and phonetic and typographical errors further challenges data matching when using name data.

The major factor that causes different name variations and errors to occur, and that determines their likely types and their distribution, is the nature of how data are being entered [72]:

- With handwritten forms or texts that are scanned and where optical character recognition (OCR) techniques are applied [133, 214], the types of error most likely to occur will be substitutions between similar looking characters (such as between 'q' and 'g' or 'S' and '5'), or substitutions of a character sequence with a single character that looks similar (such as 'm' and 'r n', or 'b' and 'l i').
- When data are typed manually, then errors can occur that are specific to the layout of the keyboard used, with neighbouring keys being hit by mistake more likely (such as 'n' rather than 'm', or 'e' instead of 'r') than keys further apart. While in certain cases this can be quickly corrected (because the resulting name or word is clearly wrong), such errors can go unnoticed due to time pressure on or distraction of the person who is doing the data entry. Spell checkers are only of limited use for personal names. Data entered through mobile devices such as tablets or smartphone will also have different error characteristics specific to the device and its error prediction capabilities.
- If data are entered through dictation over the telephone, for example through a survey study, then the dictation process is a confounding factor to the manual keyboard based data entry. If there are ambiguities with a name, the person who is doing the data entry might not request a spelling clarification or correction but rather assume a default spelling which is based on their knowledge and cultural background. Studies have shown that errors occur more likely for names that come from a language or culture that is different to the one of the person who is doing the data entry, or if names are long or complicated, such as for example 'Kyzwieslowski' [113].
- A limitation in the maximum length of characters allowed in an input field can force the use of abbreviations, initials only, or even result in disregard of certain name parts (such as middle names).
- As a final source of variations, individuals from time to time report their names in different forms, depending upon the person or organisation they are in contact with, or they deliberately provide wrong or modified names. This is commonly the case in databases that are collecting crime and fraud related information, as was discussed in Sect. 1.4.4. And while an individual might report their details accurately and consistently and in good faith, somebody else might report a family member's or friend's details inconsistently or wrongly either for malicious reasons or simply because they do not know the correct details (for example only know a person by their nickname).

For all the reasons described so far, in many situations it is not straightforward to find the 'correct' variation of a name value that is misspelt or that contains mistakes. Within the domain of data matching, one therefore has to deal with legitimate name variations as well as errors introduced during data entry and recording. While the former need to be preserved to improve data matching quality, the latter should be corrected if possible [40]. The challenge lies in distinguishing between the two. In the following four sections, different techniques for data pre-processing of names and other personal details are presented. The objective of these techniques is to convert the raw input data into a form that facilitates efficient and accurate data matching.

3.4 General Data Cleaning Tasks

Before discussing the specific steps of data pre-processing for data matching and deduplication in Sect. 3.5, in this section the three main tasks that are involved in data cleaning for any type of data analysis, mining, or processing, are presented. They are (1) handling missing values, (2) smoothing noisy values, and (3) identifying and correcting inconsistent values. Here, these three tasks are discussed with regard to their application to data matching and deduplication.

In applications such as data mining, where the aim is to detect novel and useful patterns in large databases [135], applying data cleaning can lead to much improved analysis results if the cleaning is conducted appropriately to the data mining techniques and algorithms employed. Missing values and noisy data such as outliers can have severe effects on both unsupervised and supervised learning tasks. Outliers can affect the results of data clustering, while missing values can lead to biased classification results or frequent patterns that include missing values and that therefore are not practically useful [135, 218].

When applied on data that are to be used for data matching or deduplication, different criteria need to be considered. Rather than detecting patterns, classes, rules or clusters in a database, data matching and deduplication are concerned with identifying individual records that refer to the same entities. Data cleaning must only modify the data in ways that support the application of data matching techniques. The following considerations for the three data cleaning tasks need to be taken into account:

- *Handling missing values.* Different options can be employed to handle missing values [135]:

  - Remove a record if it contains missing values. For data matching, this option will result in the removed records not being considered in the matching process at all, thereby potentially missing true matches. This option however might have to be taken if several crucial attribute values are missing in a certain record. For example, if all name and address values are missing then it is unlikely that there is enough information in other attributes to allow accurate matching.
  - Remove an attribute that contains missing values altogether from an input database, or do not use it for matching. For this option, if the attribute that contains missing values is crucial for the matching, then not considering it might be detrimental to matching quality. Even if many records have a missing value in such an attribute, then for those records that do have a value in this attribute the value should be used.

- Filling in a missing attribute value manually. This option might be possible for small databases or individual records, but this approach generally requires domain knowledge and potentially external reference data in order to identify the most likely value that should be inserted manually.
- Filling in a missing value automatically with a constant value. This option can only be applied on attributes that contain numerical values, which are rarely used for data matching or deduplication.
- Filling in a missing value with the attribute mean, median or mode. This option can only be applied on attributes that contain numerical values.
- Filling in a missing value with the mean, median or mode of a certain class of records for this attribute (for example, calculate and fill in the average salary separately for records that have a male gender from those that have a female gender). Again, this option can only be applied on attributes that contain numerical values.
- Determine the most likely value to be filled in using a rule or classification based approach. This approach is commonly used for data matching. The dependencies between certain groups of attributes allows this approach to be carried out with high efficiency and accuracy. For example, a missing gender value can be inferred based on a given name that is uniquely male or female, such as 'John' or 'Mary'. For other given names, such as 'Ashley', the gender might not be so easily determined. Another example where missing values can be inferred automatically are postcodes and suburb (or town) names. The postal services in many countries publish look-up tables of all combinations of postcodes and suburb names, and if one of these values is missing in a record and there is a one-to-one correspondence between a postcode and a suburb name then the correct value can be inferred from such look-up tables. These look-up tables can also be useful to detect and correct inconsistent values within a single record as will be discussed below.

Work on data editing and imputation has been pioneered by statisticians [107], and rule-based techniques to find the optimal value to be filled into a missing attribute value are commonly used by national census agencies to improve the quality of their survey data [143].

- *Smoothing noisy values.* Noisy data can consist of random errors or variance in values, or of outliers outside of an expected range of values (such as an age value of more than 120). They are often handled through binning, regression or clustering approaches that group similar values together and replace them by a central value such as a bin average or median, or a cluster centroid [135].

Such approaches might not be suitable when data are cleaned for data matching, because such a smoothing could result in many records having the same values in a smoothed attribute. If the values in an age attribute, for example, are binned into decades (i.e. all records with an age value from 0 to 9, 10 to 19, 20 to 29 and so on are replaced with their corresponding bin averages of 4.5, 14.5, or 24.5, respectively), then the age attribute would lose much of the discriminating information that helps identify individuals that have the same age.

Even outliers can contain information that is relevant to data matching. Returning to the example of an age value of 120 given before, if this age is based on a recorded date of birth, for example 21/07/1891, then this could potentially be a data entry error where the actual date could be 21/07/1981. Such data entry mistakes can be handled by approximate comparison techniques as will be discussed in Chap. 5.

The standardisation of attribute values described in the following section can be seen as a form of smoothing data, but applied specifically to the values in attributes that contain names, addresses, or dates, for example.

- *Identifying and correcting inconsistent values*. Here, inconsistencies within a single record and between different records need to be distinguished. The former case can sometimes be dealt with through external look-up tables and rules that (similar to filling in missing attribute values) can be used to detect if the values in two attributes contradict each other (for example a record with given name 'Paul' and gender 'F'). If such inconsistencies should be corrected or not depends upon the data at hand, and any knowledge about the quality of the data and the way they were entered. Section 3.5.4 further discusses this issue in the context of verifying the consistency of addresses.

If the attribute values in a single record are inconsistent, then at least one value needs to be changed (corrected). Unless there is certainty about which of two (or more) values is most likely the wrong one, any such change can result in further mistakes being introduced rather than corrected. In the above example, either it is assumed that the 'F' gender value is wrong and should be changed into 'M', or the given name value could be wrong and its correct value is actually 'Paula'.

Because a major aspect of the steps involved in data matching is to be able to deal with inconsistencies between attribute values, appropriate advice is to only change inconsistent attribute values if there is certainty about which value is wrong and needs to be corrected. If it is not possible to ascertain this, then the inconsistent values should rather be kept, and appropriate approximate comparison and classification techniques need to be applied that can deal with such inconsistencies but still achieve high matching accuracy. Such techniques are discussed in Chaps. 5 and 6.

Inconsistencies between different records should be corrected as much as possible before data matching or deduplication is conducted. Different codings for the same attribute, for example, either within a single database or across two databases, should be converted into the same values. For a gender attribute, for example, if one database uses the values 'F' and 'M' while the other database uses '1' and '0' then in a pre-processing step the values in either database need to be changed. Ideally, values should be changed such that they become more easily to understand and interpret [19].

Data exploration and profiling, supported through a variety of tools [19, 62, 278], are important steps that help to establish the quality of the data at hand, and to decide what types of data cleaning to employ on which parts of the data. Exploration and profiling involves collecting basic summary statistics for all attributes in a database, such as the minimum and the maximum values in an attribute, the most commonly occurring values, the distribution of the occurrence of all values in an attribute, how many records have a missing value in an attribute, and so on.

3.5 Data Pre-Processing for Data Matching

Data pre-processing refers to the tasks of converting the raw input data from the databases to be matched or deduplicated into a format that allows efficient and accurate matching [76]. Figure 3.1 on p. 42 illustrated this process on a single example record. The example databases used in the previous chapter also illustrated the process. Figure 2.2 on p. 25 shows the raw input databases, and Fig. 2.3 on p. 27 their pre-processed versions.

It is assumed that the attributes (or fields) in the input database(s) contain values that are separated by whitespace characters. These values are known as tokens. They can be words, single characters (such as initials), numbers, or compound elements such as apartment and street numbers concatenated by a slash ('3/42'), or telephone numbers made of concatenated groups of digits ('045-768-2231'). How these tokens are pre-processed is described in the following subsections.

3.5.1 Removing Unwanted Characters and Tokens

This first step of data pre-processing corresponds to a data cleaning step. The attribute values in the input database(s) might contain certain individual characters, and certain words, terms, or abbreviations, that do not contain information that is of use for data matching or deduplication, and that can and should be removed from the attribute values. Other characters or tokens need to be converted into a standardised form, for example different types of parenthesis or quotes should be replaced with one specific parenthesis or quote character, which will facilitate the standardisation and segmentation applied to the cleaned attribute values in the next steps.

Either hard-coded rules or look-up tables, such as the example shown in Fig. 3.2, are used to accomplish this first data pre-processing task. Look-up tables are generally easier to adjust to changing data needs compared to hard-coded rules, however employing hard-coded rules can be more efficient and faster than using look-up tables. For each record in the input database(s), its attribute values are scanned to see if they contain any of the tokens that are to be removed or converted. If such a token is found it is removed or converted. It is possible to have different look-up tables or rules for different types of input attributes, for example one look-up table for name attributes and one for address attributes.

Another component of this first pre-processing task is to convert all letters into either lowercase or uppercase characters, and to convert Unicode characters into ASCII characters or the other way around. Which format is chosen depends upon the characteristics of the data at hand and the limitations and requirements of the

```
# Remove characters and words from input
  ' ' := '.', '?', '~', ':', ';', '^', '=', ' na ', ' n/a '
  ' ' := ' n.a. ', ' c/o ', ' c/- ', ' also ', ' name ', '!'
  ' ' := ' only ', ' abbrev ', ' locked ', ' on ', ' of '
  ' ' := ' unk ', ' unkn ', ' missing ', '*'

# Correct words and symbols
    ' roman catholic ' := ' r/c ', ' r / c ', ' rc '
  ' church of england ' := ' c/e ', ' c / e ', ' c of e '
   ' no fixed address ' := ' nfa ', 'n/f/a ', ' n.f.a.'
      ' nursing home ' := ' n / home '
    ' other territory ' := ' o/t ', ' o.t.'
              ' and ' := '+', '&'
                ' ( ' := '<', '(', '[', '{'
                ' ) ' := '>', ')', ']', '}'
                ' | ' := '"', '"', ''', '||', '|', "'''"
                ' - ' := '-', '_'

# Correct roman numbers
              ' 1 ' := ' i '
              ' 2 ' := ' ii '
              ' 3 ' := ' iii '
              ' 4 ' := ' iv '
              ' 5 ' := ' v '
              ' 6 ' := ' vi '
              ' 7 ' := ' vii '
              ' 8 ' := ' viii '
              ' 9 ' := ' ix '
             ' 10 ' := ' x '

# Correct ordinal numbers
          ' first ' := ' 1st '
         ' second ' := ' 2nd '
          ' third ' := ' 3rd '
         ' fourth ' := ' 4th '
          ' fifth ' := ' 5th '
          ' sixth ' := ' 6th '
        ' seventh ' := ' 7th '
         ' eighth ' := ' 8th '
          ' ninth ' := ' 9th '
          ' tenth ' := ' 10th '
```

Fig. 3.2 An example correction look-up table as used by the FEBRL [62] system (described in more detail in Sect. 10.2.4). The correction works by replacing any character sequence (string in quotes) found in an attribute value of an input record that is listed on the right-hand side of a ':=' with the character sequence on the left-hand side of the ':='. As can be seen, a variety of characters and words are replaced by a single whitespace character (i.e. they are removed from an attribute value), while several variations of the same abbreviations or characters are replaced by an expanded or standardised version. Lines starting with a '#' character are comment lines

data matching or deduplication system used. A last component in this first task is to replace all multiple occurrences of whitespace characters with a single whitespace only, and to remove all leading and trailing whitespaces. For example, assuming a '␣'

symbolises a single whitespace character, the input string '␣Paul␣Peter␣␣␣Miller␣' would be converted into 'paul␣peter␣miller'.

3.5.2 Standardisation and Tokenisation

The second step of data pre-processing is the standardisation of the tokens in the attribute values by detecting and correcting values that contain known typographical errors or variations, expanding abbreviations and replacing them with standard forms, and replacing nicknames with their proper name forms.

In this data pre-processing task, individual or groups of tokens are compared with extensive look-up tables that contain values with variations and errors and their corresponding standardised and corrected values. For addresses, for example, separate such look-up tables are required for street names, locality names, state and territory names, and country names; while for personal names look-up tables are needed for title words, given names (ideally separate for female and male) and surnames. Figure 3.3 shows such a table for locality names (suburbs, towns and cities).

Such look-up tables can either be generated from databases within an organisation, or be acquired from commercial providers, or (in the cases of address data) can be available from national postal services. Look-up tables that contain common typographical variations and errors, such as the examples shown in Fig. 3.3, can be compiled from attribute values as they are entered into a database and flagged as being an unknown value. An approximate string comparison function (which will be discussed in Chap. 5) can for example be used to detect the correct value in a certain attribute that is most similar to an unknown value. Candidate variations for a look-up table can then be generated automatically to be validated by a domain expert before being added into a tagging look-up table. For example, using the look-up table in Fig. 3.3, if an input locality name 'bewerly hills' is entered by a client in a Web form, the most similar valid locality name would be 'beverly hills', and therefore 'bewerly hills' can be added as a possible variation of 'beverly hills' into the locality name look-up table.

In the tokenisation process, each token is commonly assigned one or more tags which designate the type of the token according to the look-up table(s) where this token was found, or based on some hard-coded rules. The outcomes of this process is illustrated in Fig. 3.4 assuming the look-up table from Fig. 3.3 is used. The tags are used in the third data pre-processing step to segment the sequence of tokens in an attribute value into their most appropriate output fields, as will be described in the next subsection.

The tokenisation process is normally started with the first set of tokens on the left of an attribute value. A sequence of one or several tokens is considered at any time. The tokenisation is conducted in a 'greedy' fashion [76], in that longer token sequences are considered first before shorter ones. If the longest token sequence in any of the look-up tables used in a tokenisation process contains l tokens (for example

```
# Locality names

tag=<LN>   # Tag for locality name words
               alexandria  := alezandria

           alfords point := alfonds point, alford point
           alfords point := alforts point, alfrods point

           beverley park := bevely park, bevelly park
           beverley park := beverley park, beverlly park

           beverly hills := beverley hills, beverly hill

           sydney airport := syd inter airport, syd airport

    the university of sydney := sydney university, sydney uni
    the university of sydney := uni sydney, university sydney
```

Fig. 3.3 An example tagging look-up table as used by the FEBRL [62] system. The tag 'LN' is used to designate all following entries in this look-up table as locality names. A sequence of tokens that occurs in an attribute value that is listed on the right-hand side of the ':=' will be replaced by the sequence of tokens on the corresponding left-hand side, and the sequence of tokens is assigned the 'LN' tag, as illustrated in Fig. 3.4

'syd inter airport' contains $l = 3$ tokens), then at any step of the process the next l tokens in an input attribute value are considered. If the tokenisation process starts from the left, then the first l tokens, denoted with $t[1], t[2], \ldots, t[l]$, are considered to be the candidate set of tokens in the first step. If these l tokens match a token sequence in any of the look-up tables, then they are replaced by the sequence of corrected tokens, and the tag of this corrected sequence is assigned to the set of tokens.

For example, using the look-up table from Fig. 3.3, if a token sequence starts with 'syd inter airport' then these three tokens are replaced by the standardised compound token 'sydney airport' which is assigned an 'LN' tag to designate that it corresponds to a known locality name. Note that even correct known token sequences that are found in an input field, such as 'sydney airport', are assigned the corresponding tag.

If at any step in the tokenisation process no token sequence of length l is found in the input attribute value, then the length of the candidate token sequence is reduced from l to $l-1$ (i.e. tokens $t[1]$ to $t[l-1]$ are considered), and again all look-up tables are searched for this token sequence. The length of the candidate token sequence is reduced until either a token sequence is found in a look-up table, or a single token is assigned an appropriate hard-coded tag (as for example listed in Table 3.1). It is also possible that a token is assigned several tags if the token is found in several look-up tables, as shown in Fig. 3.4.

It is important that longer candidate token sequences are considered first, such that for example the token sequence 'sydney uni' is correctly identified to correspond to the standardised locality name 'the university of sydney', rather than the single

Paul	Peter	Miller
paul	peter	miller
GM	GM,SN	SN

17	Epinng	Rd	Bevely Park	N.S.W.	2011
17	epping	road	beverly park	nsw	2011
NU	LN,SN	ST	LN	TR	PC

Fig. 3.4 Two examples of input values, the first being a name and the second an address. The first row in each example shows the raw uncleaned input, the second row shows the cleaned and standardised tokens, and the third row shows the tag(s) assigned to each token that indicate their type. A description of these tags is given in Table 3.1

token 'sydney' is assigned as locality name and then the second token 'uni' is left as a potentially unknown token.

The standardisation and tokenisation process continues as long as there are unprocessed tokens in an attribute value. At the end of the tokenisation process, all tokens in an attribute value will have been replaced by corrected and standardised forms, and they will have one or more tags assigned to them, as illustrated in Fig. 3.4.

3.5.3 Segmentation into Output Fields

The third step of data pre-processing is the segmentation of the tokenised and tagged attribute values into well-defined output fields that are suitable for data matching or deduplication, as was illustrated in Fig. 3.1. This step is the most challenging step in data pre-processing, because often there are several possible assignments of tokens to output fields. The challenge is to identify the most likely assignment. This task is also known as *parsing* [143], and is related to the field of *information extraction* which is concerned with identifying structured information in semi-structured or free format text [230].

The objective of segmentation is to have each output field contain a single piece of information, made of one or a small number of tokens, rather than having several pieces of information in one field or attribute, as was illustrated in Fig. 3.1 on p. 42. The values in these output fields are then used in the detailed pairwise comparison of record pairs (as will be discussed in Chap. 5), which generally leads to much improved matching quality compared to when the unstandardised and unsegmented input attribute values would be used. The following lists show the output fields that are commonly used for data matching or deduplication:

- *Personal names.* Title, name prefix, given or first name, initials, middle name, family name or surname, alternative family name or surname, name suffix.

- *Street addresses.* Unit prefix, unit type, unit number, unit suffix, street or wayfare number, street or wayfare name, street or wayfare type, building name, postal address number, postal address type, institution name, institution type.
- *Address localities.* Locality or town name, territory or state name, postcode or zipcode, country.
- *Dates.* Day, month, year.
- *Telephone numbers.* Country code, area code, number, extension.

Not all of these output fields will be available in all databases, and for many records some of the fields will not contain a value. The actual output fields used and their names also depends upon the data at hand and of course will differ from country to country.

Challenges occur when there are ambiguities in a token sequence that is to be segmented. For example, the middle name 'Peter' in the three name words 'Paul Peter Miller' shown in Fig. 3.4 could either refer to this person's middle name or to his surname (with 'Miller' being a second surname from the original compound surname 'Peter-Miller').

Different segmentation techniques have been developed for different types of input data, such as personal names, business names, or addresses. Sections 3.6 and 3.7 will cover the two main types of techniques employed, rule-based and statistical, in more detail.

The standardisation and segmentation steps in data pre-processing are not necessarily independent of each other. They can be combined into one process, where the segmentation is conducted on the unstandardised tokens first, and the standardisation is applied based on the segmented tokens. For example, the abbreviation 'St' in an address attribute can either stand for the street type word 'Street', or be part of a town name such as 'Saint Mary', depending on the overall token sequence in the address.

3.5.4 Verification

A possible fourth step of data pre-processing is the verification of the correctness of the values assigned to the different output fields, and the validation of value combinations in several attributes. For names, for example, such verification can include checking if the combination of values in the given name and gender attributes are valid (a given name 'John' and gender value 'F' is generally not a valid combination), or if a title word does contradict the given name value of a record (a title 'Ms' is not valid for a record with given name 'John'). Such tests can be based on look-up tables of known give names that are uniquely male or female, as was previously discussed in Sect. 3.4.

For addresses, testing the existence and correctness of address values can be carried out using external reference databases, that, for example, contain all validated addresses in a country. Such reference databases are commonly available from national postal services or commercial providers. They allow the verification of

Table 3.1 List of tags used by the data standardisation module of the FEBRL system [62]

Tag	Description	Component	Based on
LQ	Locality qualifier word	Address	Look-up table
LN	Locality (town, suburb) name	Address	Look-up table
TR	Territory (state, region) name	Address	Look-up table
CR	Country name	Address	Look-up table
IT	Institution type	Address	Look-up table
IN	Institution name	Address	Look-up table
PA	Postal address type	Address	Look-up table
PC	Postcode (zipcode)	Address	Look-up table
N4	Numbers with four digits (not known postcodes)	Address	Hard-coded rule
UT	Unit type (e.g. 'flat' or 'apartment')	Address	Look-up table
WN	Wayfare (street) name	Address	Look-up table
WT	Wayfare (street) type (e.g. 'road' or 'place')	Address	Look-up table
TI	Title word (e.g. 'ms', 'mrs', 'mr', 'dr')	Name	Look-up table
SN	Surname	Name	Look-up table
GF	Female given name	Name	Look-up table
GM	Male given name	Name	Look-up table
PR	Name prefix	Name	Look-up table
SP	Name separators and qualifiers (e.g. 'aka' or 'and')	Name	Look-up table
BO	'baby of' and similar values	Name	Look-up table
NE	'nee', 'born as' or similar values	Name	Look-up table
II	Initials (one letter token)	Name	Hard-coded rule
ST	Saint names (e.g. 'saint george' or 'san angelo')	Address/name	Look-up table
CO	Comma, semi-colon, colon	Address/name	Hard-coded rule
SL	Slash '/' and back-slash '\'	Address/name	Hard-coded rule
NU	Other numbers	Address/name	Hard-coded rule
AN	Alphanumeric tokens	Address/name	Hard-coded rule
VB	Brackets, braces, quotes	Address/name	Hard-coded rule
HY	Hyphen '–'	Address/name	Hard-coded rule
RU	Rubbish (for tokens to be removed)	Address/name	Look-up table
UN	Unknown (none of the above)	Address/name	Hard-coded rule

This table is adapted from Table 3 in [76]. A 'hard-coded rule' refers to the cases where a specific piece of program code is used to assign a tag to an input character or token. As can be seen, most tags are based on look-up tables and specific to either addresses or names. Some of the hard-coded tags take care of special characters or help to characterise tokens not found in any of the look-up tables

different parts of a segmented address, including the verification of locality name and postcode combinations, and if such a combination is known in a given territory or state. Other verification steps for addresses include the test if a street name and type combination occurs in the locality value given in a record, and even if a street number occurs in the given street or not.

If an invalid combination is found then it can either be flagged for manual inspection, or be corrected automatically (but being aware of the potential that a correction can introduce new errors, as was described on p. 51). In case no correction is being made for an invalid combination, a flag can be added to the record indicating the attributes that contain inconsistent values. This information can then be used in the matching process to, for example, lower the similarity value between two records if the flag indicates that some of the address values in a record might be wrong.

3.6 Rule-Based Segmentation Approaches

Rule-based techniques for segmentation of names and addresses have been employed in the field of data matching for several decades. The basic idea of such techniques is to process the list of tokens and tags either from left to right or from right to left, and using hand-crafted or learned rules to assign the token or tokens covered by a rule to their appropriated output field.

Processing token sequences starting from the left is appropriate for most name values from Western countries, as well as the street component of addresses, while processing token sequences starting from the right can be appropriate when locality details (postcodes, suburb, state, and country names) are available in the attributes to be standardised.

Rule-based approaches are best suited for input fields that contain controlled and well-structured information, such as telephone numbers or names that are made of only a small number of tokens [230]. For addresses, developing efficient and accurate rule-based systems is much more difficult [76], because a much larger number of rules is needed that can deal with the much larger variability in token sequences that represent addresses.

A rule-based system is made of two parts. The first is a set of rules in the form of 'if *condition* then *action*' [217]. The *condition* of a rule tests for the occurrence of a certain tag or tag sequence, and the *action* is the assignment of the tokens covered by a rule into the appropriate output fields. The *condition* of a rule is generally testing for tags rather than tokens, because tags are more general than tokens and therefore rules based on tags can cover more variability in the input. If the *condition* of a rule is true then the rule is 'triggered' or 'fired' and the *action* of the rule is executed.

The second part of a rule-based system is the ordering or the policies of which rules should be fired first when the *condition*'s of several rules are true for a certain sequence of tags. The ordering can either be based on the specificity of the rules, in that rules that cover more tags are fired first, or it can be based on which output fields are most important and should have values assigned to them (for example, the given name and surname output fields are more important than the middle name or name suffix fields), or the ordering can be based on a manual sorting of the rules using domain knowledge. Often various heuristics are applied, and special cases are handled with individuals rules [230].

Figure 3.5 shows an example subset of rules for segmenting name values. Rules are normally applied on the tags (denoted with $t[i]$) that have been assigned to the tokens (denoted with $o[i]$) in the tokenisation step as was described in Sect. 3.5.2. While the eight rules shown may cover most known simple names in a database, more complex

---

if t[i] = 'TI' **then** title ← o[i]
if t[i] = 'PR' **then** name_prefix ← o[i]

if t[i] = 'GM' and t[i+1] = 'SN' **then** given_name ← o[i], surname ← o[i+1]
if t[i] = 'GF' and t[i+1] = 'SN' **then** given_name ← o[i], surname ← o[i+1]
if t[i] = 'SN' and t[i+1] = 'GM' **then** given_name ← o[i+1], surname ← o[i]
if t[i] = 'SN' and t[i+1] = 'GF' **then** given_name ← o[i+1], surname ← o[i]

if t[i] = 'UN' and t[i+1] = 'SN' and $i + 1 = L$ **then** given_name ← o[i], surname ← o[i+1]
if t[i] = 'SN' and t[i+1] = 'UN' and $i + 1 = L$ **then** given_name ← o[i+1], surname ← o[i]

Fig. 3.5 A small example of a subset of rules used to segment a name input value using the tags defined in Table 3.1. The sequence of tokens is denoted by $o[i]$ and the corresponding sequence of tags by $t[i]$, with $1 \le i \le L$ and L being the number of tokens in the given name input value. The first rule assigns a known title token into the *title* output field, while the second rule assigns a known name prefix into its appropriate output field. The next four rules assign known given name and surname values into the appropriate fields, while the last two rules only assign given name and surname values into the corresponding output fields if there is no other tag that follows. Only one rule is applied on a tag (or tag sequence), and once the tag (or tag sequence) is covered by a rule it is removed from the input tag sequence

names made of several components (name prefixes and suffixes, middle names, etc.) will require many more specific rules.

When a rule-based system is developed based on hand-crafted rules, initially the basic rules (such as the ones shown in Fig. 3.5) are implemented and applied on the data that are to be standardised. All records that are not covered by any rule are then used to develop additional rules. Such a *failure driven* iterative approach [217] over time results in a rule-base that covers most if not all variations of tag sequences that occur in an attribute. The manual investigation of the input values not covered by any rule, and generating appropriate new rules for them, is however a labour intensive task that needs to be repeated each time data with new characteristics are to be standardised.

There are various ways of how rules can be represented, including regular expressions, SQL statements, pattern items and lists, and even specific pattern languages or scripts written in programming languages such as Java or C++ [230]. The early AutoStan/AutoMatch [251] suite of data cleaning and matching software, developed in the 1990s by Matthew Jaro (formerly of the US Census Bureau and founder of MatchWare Technologies), for example, employed a look-up table based tokenisation phase followed by a re-entrant regular-expression rule-based parsing and segmentation phase. Regular expressions allow for rules where tokens are checked for certain string patterns (not just equality). For example, the test **if** o[i] = 'stre?t' **then**. . . will return true for the two words 'stret' and 'street'. The re-entrant approach of AutoStan means that once a token (or token sequence) is covered by a rule (and assigned into an output field), the token (or token sequence) is removed from the input token sequence and the rule-base is applied on the new shorter token sequence. AutoStan rule-bases employed for segmenting addresses could contain hundreds if not thousands of rules for production systems.

An alternative to the labour intensive manual development of a rule-base is to employ a rule-learning algorithm [230]. Rules can be learnt in an automatic fashion if training data in the form of correctly segmented input examples are available. Such training data can either be generated manually, or be the output of an earlier segmentation of similar data. In its most general form, a rule learning system learns individual rules that are disjunctions and that each cover a subset of the training data set.

Assuming a training data set D is provided that contains n training records, $d_1, \ldots, d_n$, each made of a sequence of tags and the output fields they are assigned to. Figure 3.6 shows three such training records. The objective of a rule learning system is to learn a set of k rules $r_1, \ldots, r_k$ that cover all training records in D. The *condition* part of each rule r covers a certain subset $s(r)$ of all training records in D, which is called the *coverage* of a rule. The *action* part of a rule will be correct for some training records in $s(r)$ and wrong for others. The subset of training records where the action is correct is denoted with $s'(r) \subseteq s(r)$. The *precision* of a rule is calculated as $p = |s'(r)|/|s(r)|$ [230].

The objective of a rule-learning system is to learn rules that provide good coverage and have high precision, because such rules will be well suited for the segmentation of new unsegmented input records. Finding the optimal set of rules for a given training data set is intractable, therefore practical rule learning algorithms are based on heuristic approaches [230]. The two broad categories of heuristics are bottom-up and top-down approaches. In the first category, a very specific rule (that only covers one training record) is made more general by removing a test from the *condition* of the rule such that the coverage of the rule is increased (at the cost of likely loosing some precision). In top-down approaches, general rules that cover many training records but have low precision are made more specific by adding further tests to the *condition* of the rule until some stopping criteria is reached. There are various ways of how these approaches can be implemented and different algorithms have been developed, including *Rapier*, $(LP)^2$, *FOIL* and *WHISK*. An excellent survey of rule-based approaches to information extraction is provided by Sarawagi [230].

3.7 Statistical Segmentation Approaches

A major drawback with rule-based approaches to data segmentation is that rules are hard, meaning that they either fire or do not fire (i.e. cover a set of tokens in an input sequence or not), depending upon if a certain condition is fulfilled [230]. In practice this means that for unseen variations in the input data that are not covered by any rule in a rule base, a new rule is required. Manually generating a comprehensive rule-base is labour intensive and time consuming, and requires adjustments each time the data to be segmented changes [217]. Techniques that learn rules from training data also require training examples for any variation in the data in order to be able to learn the rule or rules that cover these examples. Therefore, comprehensive training data sets are required, which again can be expensive to generate or collect.

17	epping	road	beverly park	nsw	2011
NU	LN	ST	LN	TR	PC
Wayfare number	Wayfare name	Wayfare type	Locality name	Territory	Postcode

	main	street	sydney	nsw	2000
	WN	ST	LN	TR	PC
	Wayfare name	Wayfare type	Locality name	Territory	Postcode

42	george	ally	newtown		2067
NU	WN	ST	LN		N4
Wayfare number	Wayfare name	Wayfare type	Locality name		Postcode

Fig. 3.6 Three example training sequences of addresses consisting of tokens (top rows in each example), tags (middle rows) and output fields (bottom rows). The tags are based on the list given in Table 3.1, while the output fields correspond to those shown in Fig. 3.1. Each token is assigned one tag only, and only the tags and names of output fields are used for training a rule-based or statistical segmentation model, but not the actual tokens that make up an address

Statistical approaches to segmentation try to overcome the rigid decision making of rule-based systems. They are instead based on probability distributions that provide likelihoods of which token in an input should be assigned to which output field. Similar to rule-based systems, these probabilities are learned from training data that consist of segmented token and tag sequences where each token is assigned its appropriated tag, as illustrated in the examples shown in Fig. 3.6. Similar to rule-based systems, tag sequences rather than token sequences are used to train the statistical segmentation models and to segment new input values [76].

The process of assigning tokens to output fields can be seen as a classification process where each token is assigned to its most likely output field (more generally called *label*) according to the learned model [230]. However, this task is following an ordering, in that the classification of a token depends upon the classification of the previous token(s) (assuming an assignment of tokens starts from the left of a token sequence) and possibly also the following token(s).

Different statistical models have been developed that capture these dependencies within tag sequences. The most popular techniques have been hidden Markov models (HMMs) [223], maximum entropy Markov models (MEMMs), and more recently Conditional Random Fields (CRFs). CRFs are capable of modeling a single joint distribution over the sequence of the predicted output fields for a given token sequence. The dependency of the classified output field of a token is based on the adjacent previous and next output fields. For HMMs, on the other hand, the classification of a token only depends upon the classification of the previous token but not the following one.

The process of training a statistical model for segmentation is based on either calculating the maximum likelihood or maximum margins for all given tag sequences in the training data. The detailed mathematical descriptions of these techniques are outside the scope of this book, the interested reader is refereed to the excellent survey given by Sarawagi [230]. In the remainder of this section, an example of an address segmentation approach based on HMMs is provided, which previously has been shown to outperform a manually developed rule-based approach [68, 76].

3.7.1 Hidden Markov Model Based Segmentation

Hidden Markov models [223] were developed in the 1960s and 1970s. They are widely used in speech recognition and natural language processing. They are computationally efficient to train and are able to handle new unknown sequences in a robust fashion. They have been employed by several researchers for name and address segmentation [41, 56, 68, 76, 240].

A HMM can be viewed as a probabilistic finite state machine that consists of a set of (hidden) states, transition links between these states, and a set of output (or observation) symbols. Each link between two states has a nonzero probability assigned with it, and each state emits output symbols with a certain probability distribution. The transition and output probabilities are stored in two matrices. A simple example of a HMM for address segmentation, together with its transition and output probability matrices, is shown in Fig. 3.7.

Two special states of a HMM are the *Start* and *End* state. Beginning with the *Start* state, a trained HMM generates a sequence of output symbols $O = o_1, o_2, \ldots, o_k$ by making $k - 1$ transitions from one state to another until the *End* state is reached. The output symbol o_i, $1 \leq i \leq k$, generated in state i, is based on this state's probability distribution of the output symbols. The *Start* and *End* states are not actually stored in a HMM because no output symbols are emitted in these states. Instead of the *Start* state a list of initial state probabilities is used that provide the likelihoods that a sequence starts with a certain state.

For a given trained HMM, it is possible that the same sequence of output symbols can be generated by taking different paths through the HMM. Each path, however, will have a different probability according to the transition probabilities between the states in the path. Given a certain sequence of output symbols, for the task of segmentation one is interested in the most likely path through a given HMM that will generate this sequence. Using a dynamic programming approach, the *Viterbi* algorithm is an efficient way to compute this most likely path for a given sequence of output symbols [223].

Training data in the form of sequences of (state name, output symbol), possible manually prepared, are required to learn the transition and output probabilities. Each training record corresponds to a path through the HMM from the *Start* to the *End* state. While the set of output symbols can be created using the training data, the states of a HMM are generally fixed and are defined before training. When segmenting addresses, for example, the set of states will correspond to all possible output fields of an address, such as the ones listed on p. 56, while the output symbols correspond to all possible tags (as for example listed in Table 3.1) that can occur with addresses.

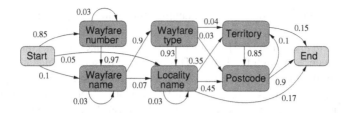

	To state						
From state	Wayfare number	Wayfare name	Wayfare type	Locality name	Territory	Postcode	*End*
Start	0.85	0.1	0.0	0.05	0.0	0.0	0.0
Wayfare number	0.03	0.97	0.0	0.0	0.0	0.0	0.0
Wayfare name	0.0	0.03	0.9	0.07	0.0	0.0	0.0
Wayfare type	0.0	0.0	0.0	0.93	0.04	0.03	0.0
Locality name	0.0	0.0	0.0	0.03	0.35	0.45	0.17
Territory	0.0	0.0	0.0	0.0	0.0	0.85	0.15
Postcode	0.0	0.0	0.0	0.0	0.1	0.0	0.9

Output symbol	State					
	Wayfare number	Wayfare name	Wayfare type	Locality name	Territory	Postcode
NU	0.9	0.01	0.01	0.01	0.01	0.05
WN	0.01	0.5	0.01	0.1	0.01	0.01
WT	0.01	0.01	0.92	0.01	0.01	0.01
LN	0.01	0.1	0.01	0.8	0.01	0.01
TR	0.01	0.06	0.01	0.01	0.93	0.01
PC	0.03	0.01	0.01	0.01	0.01	0.8
N4	0.02	0.01	0.01	0.01	0.01	0.1
UN	0.01	0.31	0.02	0.05	0.01	0.01

Fig. 3.7 A simplified Hidden Markov model (top) for addresses, based on the output fields shown in Fig. 3.1. The table in the middle shows the transition probabilities, while the table at the bottom shows the output probabilities. Adapted from [76]

The training process iterates over all training records and adjusts the transition and output probabilities according their output field and tag sequences. For example, the transition probability of 0.93 from state 'Wayfare type' to 'Locality name' in Fig. 3.7 results from 93 % of all training records containing these two output fields in sequence, while only 4 % of training records had a value in the 'Territory' output field directly after a value in the 'Wayfare type' output field, and only 3 % of training records had a 'Wayfare type' output field that was directly followed by a 'Postcode' output field.

Instead of using the actual tokens found in an input field, the tag or tags that were assigned to each token are used as the output symbols of the HMM [76]. This makes a HMM more general and more robust, and also computationally more efficient because the number of different tags is much smaller than the number of different tokens that will be encountered in the input data.

Once a HMM is trained on a set of example input values, sequences of tags from new records that are to be segmented into output fields can be segmented efficiently using the *Viterbi* algorithm [223], which returns the most likely state sequence of the given tag sequence through the HMM. This sequence of states, which corresponds to a sequence of output fields, is then used to assign each token of an input value to an output field. For example, consider the following token sequence from an address with its corresponding tag sequence which is based on the tags from Table 3.1 on p. 57:

32	Garden	Place	Brisbane	7014	Queensland
NU	WN	WT	LN	PC	TR

Applying the Viterbi algorithm for this tag sequence on the example HMM from Fig. 3.7 will lead to the following state sequence which has the highest likelihood:

Start → Wayfare number → Wayfare name → Wayfare type → Locality name → Postcode → Territory → *End*.

The tokens in this address will therefore be assigned to the following output fields:

Wayfare number:	32
Wayfare name:	Garden
Wayfare type:	Place
Locality name:	Brisbane
Territory:	Queensland
Postcode:	7014

When generating models for segmentation of address and other types of data, a major issue in many application domains is how to collect appropriate training data. Such training data need to be of high quality and be broad enough to cover the diversity of input values that likely occur in the attributes that are to be segmented.

One possible approach is to bootstrap the training process by manually cleaning, tokenising and segmenting a small number of input values and assigning each tag to its most likely output field (such as the examples shown in Fig. 3.6), to then use this small set of training data to train a first segmentation model (such as a first rough HMM), and to then use this first model to segment a larger number of input values [76, 230]. This second set of segmented input values will likely contain too many wrongly segmented values, and careful manual inspection and correction of these input values is required. Once done, a second training set of segmented input values is available that can be used to train a second HMM. This second HMM will likely be more accurate than the first one. This process of segmenting input values using a HMM, correcting the wrongly segmented values, and using the new set of segmented input values to train a more accurate HMM can be repeated until a HMM of satisfactory quality is available. This approach has shown to be much less time consuming compared to the manual generation of hand-crafted rules [76].

An alternative approach is to use cleaned and segmented input values that are available either in a reference databases or from earlier segmentation of the same types of data [5, 65]. The important aspect with this approach is that these data are of high quality and contain a large diversity of attribute values, such that the trained segmentation model is robust with regard to different unknown input addresses. For addresses, such reference databases can either be obtained from national postal service or they are already available in a database or data warehouse of an organisation. The structure and attributes of these segmented addresses needs to be the same as the desired structure of the addresses that are to be segmented. Having access to such a reference database allows an automatic learning process of segmentation models which can provide fully automated address standardisation [65].

3.8 Practical Considerations and Research Issues

An important initial activity in any data matching or deduplication project must be the assessment of the quality of the data that are to be matched. Known as data exploration or data profiling, this task can be achieved through a variety of tools that are either integrated in a data matching software, are external standalone programs, or are part of larger data processing, analysis or data warehousing systems.

At a minimum, for each attribute that will be used for matching, the number of different attribute values and their frequency distribution, the type of values in an attribute (such as string, number, date, etc.), as well as the number of records that have an empty value in an attribute should be known. This information is relevant when attributes are selected to be part of blocking keys during the indexing step (as will be discussed in the following chapter), and when appropriate comparison functions are chosen in the comparison step (as was covered in Chap. 5).

Many data cleaning and standardisation techniques rely heavily upon look-up tables. These tables contain, for example, personal names and their variations and common misspellings, or suburb, town, or state names, and postcodes from a certain country. To achieve cleaned and standardised data that are of high quality, it is important that these look-up tables are carefully customised according to where the data to be matched are sourced from. This does not just hold for names used in addresses, but also for given names and surnames which often have different spelling variations in different countries (even for example within English speaking countries). While such customisation of look-up tables will initially be a time consuming and labour intensive process, in the end the effort will be worthwhile because of the improved matching quality that can be achieved. Further on, the cleaned and standardised data will likely be useful for other applications within an organisation as well. Besides look-up tables, both the rules in rule-based segmentation systems and the training data for statistical segmentation systems also need to be customised to the data that are being matched or deduplicated.

What type of data cleaning and standardisation approach to use depends both upon the quality of the raw input data, and the amount of resources (with regard to labour, funding, and computing power) that is available for a given data matching or deduplication project. A further practical consideration is if a matching or dedupli-cation exercise on a certain set of data is a one-off project or if it is likely that the data will be reused for future data matching projects. In the latter case it is worth to invest more efforts into data pre-processing than in the first case, especially if the matched data are used as an authoritative data repository (such as a master patient index database) for different applications within an organisation, and any new data will be matched with this authoritative data repository.

While data quality has been recognised as a massive problem that costs many organisations large amounts in lost revenue and wasted resources, the amount of research in the area of data pre-processing (cleaning and standardisation) is surpris-ingly low. One reason for this might be that data pre-processing is a very domain specific task that involves significant amounts of domain expertise and manual cus-tomisation and intervention. How to automate data pre-processing techniques with the aim to reduce manual efforts will be a valuable research undertaking.

Another interesting research direction will be to investigate how well different data pre-processing techniques are able to improve the outcomes of matching or deduplicating different types of data, and if there is a way to identify an optimal approach to how data pre-processing should be applied. This question can only be considered in combination with a specific data matching technique employed. Still, a large comparative investigation of different data cleaning and standardisation techniques applied on databases of different quality and with different characteristics would lead to a much improved understanding of how data pre-processing affects the outcomes of a data matching or deduplication exercise.

3.9 Further Reading

There is a large body of work available on the topic of data quality, addressing the many issues and challenges involved in this topic from different angles. Batini and Scannapieco [19] provide a detailed discussion of concepts, methodologies and techniques that can be employed to assess and improve data quality. Pyle [218] covers data quality and data preprocessing specifically for data mining applications. Lee et al. [177] on the other hand provide a road map to data quality that covers this topic at a less technical level more suitable for managers and practitioners that need to implement systems where data quality is important.

The many different issues that can arise when dealing with names have been discussed by various authors [40, 57, 72, 175, 208, 210, 243]. A large body of infor-mation about names is also available in online resources that cover names and their origins, names and their variations, and the changing popularity of baby names.

The interested reader is referred to Web sites such as: http://www.thinkbabynames.com, http://www.babynames.com, http://www.rogerdarlington.co.uk/useofnames.html, and http://en.wikipedia.org/wiki/Personal_name.

An excellent recent survey of information extraction techniques which is of relevance to name and address segmentation is provided by Sarawagi [230]. Techniques that specifically deal with data cleaning and standardisation for data matching are presented by Churches et al. [76] and by Herzog et al. [143]. The use of reference databases to automate the standardisation process of addresses has been described by Agichtein and Ganti [5] and Christen and Belagic [65].

Two novel approaches to data cleaning have recently proposed by Arasu and Kaushik [12] who used a grammar-based framework that can be used to reason about and manipulate data representations, and Guo et al. [130] who employed latent semantic association to conduct unsupervised address standardisation.

Chapter 4
Indexing

4.1 Why Indexing?

The simple example given in Chap. 2, specifically Figs. 2.4, 2.5 and 2.6 on pp. 29 and 31, helps to illustrate that even when matching small databases the majority of comparisons between records will correspond to non-matches. These are comparisons between two records that each refers to a different entity. As will be covered in Chap. 5, the detailed comparison of records can be a computationally expensive undertaking, with some comparison functions having a computation complexity that is quadratic in the lengths of the attribute values (that most commonly are strings) that are compared. The comparison step is generally the computationally most expensive step in the data matching process.

The aim of indexing in data matching is to reduce the number of record pairs that are compared in detail as much as possible, by removing pairs that unlikely correspond to true matches. At the same time, all record pairs that possibly correspond to true matches (i.e. where the two records of a pair refer to the same entity) need to be kept for detailed comparison. Without indexing, the matching of two databases that contain m and n records, respectively, would result in $m \times n$ detailed record pair comparisons. For large databases, this is clearly not feasible.

Indexing can be seen as a filtering or searching step. Because it is mostly based on some form of index data structure that brings 'similar' values together (what is similar will be discussed below), the term *indexing* is commonly used to name this step of the data matching process [64].

The general approach of indexing techniques is to process all records of the databases to be matched and to either insert each record into one or several blocks, lists or clusters, according to some criteria, or to sort the databases such that similar records are moved closely together. The criteria used is commonly called a 'blocking key' (the term used in this book) or 'sorting key'. The blocking or sorting key values are generated based on the values of either a single or from several attributes. These values are often encoded, as will be discussed below. As an example blocking key, a postcode (or zipcode) attribute could be used, such that all records that have the same

P. Christen, *Data Matching*, Data-Centric Systems and Applications,
DOI: 10.1007/978-3-642-31164-2_4, © Springer-Verlag Berlin Heidelberg 2012

postcode value will be inserted into the same block or index list. For a sorting-based indexing technique, using this sorting key the databases will be sorted according to postcode values. This sorting will lead to records that have the same value in the postcode attribute being next to each other.

A critical aspect for any indexing technique is the definition of the blocking key or keys used. As will be further discussed in the following section, a major consideration is the quality of the values in the attributes used as blocking keys, especially their completeness (how many records have values in an attribute), and the frequency distribution of the values in these attributes. Both these characteristics affect the number of candidate record pairs that are generated, and their quality (i.e. how many refer to true matches or not). This in turn will affect the overall accuracy and completeness results of a data matching exercise.

Indexing is not just important for data matching, it also needs to be applied for the deduplication of a single database. Without indexing, in a deduplication project each record in a database would be compared with all others, resulting in a total of $n \times (n-1)/2$ record pair comparisons for a database that contains n records. Because the comparison of two records is symmetric, each pair only needs to be compared once. The indexing techniques used for the matching of two databases can also be applied for the deduplication of a single database.

The question of how to evaluate indexing techniques will be discussed in detail in Sect. 7.3. Three measures are generally used [71]. The first measure, known as reduction ratio, calculates how many candidate record pairs are generated by an indexing technique compared to all possible record pairs (full naive pair-wise comparison of all pairs). The second measure, called pairs completeness, calculates how many record pairs that refer to known true matches are included in the candidate record pairs (this measure corresponds to recall as used in information retrieval [288]). The third measure, pairs quality, calculates how many of the candidate record pairs generated correspond to true matches (this measure corresponds to precision as used in information retrieval). The last two measures require knowledge about the true match status of all record pairs, which is commonly not available in real-world matching or deduplication situations (a topic which will be covered in detail in Chap. 7).

4.2 Defining Blocking Keys

As a recent experimental survey highlighted [64], one of the most important aspects of the indexing step is not which indexing technique is employed in a data matching project, but the definition of the blocking key(s) that results in similar records being successfully grouped into the same block(s).

At this point a discussion about what constitutes 'similar records' is warranted. Depending upon the data to be matched, similarity between attribute values can refer to phonetic similarity (for example how similar two names sound), character shape similarity (for example how similar two written names look) or numerical similarity (for example how close two age or date values are to each other). If it is known how the data in the databases to be matched or deduplicated were recorded or entered, then an appropriate encoding function can be applied on attribute values when the blocking key values (BKVs) are generated from them. The aim of such an encoding is to bring similar values together such that they are inserted into the same block.

The most common forms of data entry are manually typed values (from hand-written or typed forms), values scanned and automatically recognised using OCR technology, or values dictated and transcribed using an automatic speech recognition system [72]. A widely used approach in indexing to convert 'similar' values into the same BKV is to employ a phonetic encoding function, such as Soundex, NYSIIS or Double-Metaphone [57]. These functions replace a (name) string with a code that reflects how a name would sound if it is spoken. Names that sounds similar are converted into the same code. The phonetic encoding functions most commonly used for data matching are presented in the following section.

When blocking keys are defined based on the attributes available in the databases to be matched or deduplicated, then several issues need to be considered.

- *Attribute data quality*: the quality of the values in an attribute will influence the quality of the BKVs generated. If an attribute has a missing or empty value in a large portion of records, then as a result many records will be added into a block where the BKV is an empty value. This raises the question of whether having an empty value in an attribute means that the records that have an empty value are similar to each other or not.

  Ideally, an attribute used to generate BKVs should be as complete as possible, i.e. all records in a database should contain a value in this attribute. These values should also be of high quality, because any error in a value that results in a different BKV will mean that a corresponding record is inserted into the wrong block [71]. For example, if a postcode value is recorded wrongly as '2130' rather than '2730' (possibly an OCR mistake), then the corresponding record might be inserted into the block with BKV '2130' rather than '2730', and therefore it would not be compared with the records that it potentially matches with.

- *Attribute value frequencies*: the frequency distribution of the values in an attribute used as part of a blocking key will influence the number of candidate record pairs that are generated. If the frequency distribution is skewed such that some values are very frequent, then these most frequent values will dominate the number of candidate pairs that are generated.

  For example, if an attribute that contains surnames is used to generate BKVs, then the two blocks generated from the surname values 'smith' and 'miller' (two of the most common surnames in many English speaking countries) will be large. If for example two databases where each contains 1 million records are being matched, and only one percent (10, 000 records) contains the surname 'smith', then this single block will generate $10,000 \times 10,000 = 100,000,000$ candidate record pairs that are to be compared in detail. This clearly would not make sense, and the information contained in other attributes of these records (such as given name, age or postcode values) would not only reduce the number of record pair comparisons, but also increase the likelihood that the compared pairs do refer to the same person.

The application of an encoding function, such as Soundex for example, can make the problem of large blocks even worse, because several attribute values are mapped into the same encoding value and therefore into the same block. This can result in more records being inserted into the same block.

It is therefore of advantage to select attributes as blocking keys where attribute values have a frequency distribution close to the uniform distribution, resulting in blocks that are of equal sizes.

- *Trade-off between number and size of blocks*: the third issue that needs to be considered is the trade-off between the number of BKVs (and thus the number of blocks) and the size of the blocks generated (and thus the number of candidate record pairs generated) [20, 64]. On the one hand, a small number of large blocks will result in a larger number of candidate record pairs that likely contain more of the true matching record pairs. On the other hand, a large number of small blocks will lead to a smaller number of candidate record pairs (and thus a reduced run time) at the cost of potentially missing more of the true matching pairs.

The more specific a blocking key definition is, the smaller the resulting blocks will become and therefore less record pair comparisons need to be conducted. A more specific blocking key definition can be achieved by concatenating values from several attributes, possibly encoded first, as illustrated in Fig. 4.1.

As will be discussed later in this chapter, some of the presented indexing techniques are more sensitive to the choice of blocking keys than others. Having a more specific blocking key that leads to a larger number of smaller blocks is also of advantage for indexing techniques that sort the databases according to the blocking key (or sorting key), because a larger number of BKVs allows a more fine-grained sorting of the records in a database. This will be described in detail in the relevant sections later in this chapter.

As was discussed in the previous chapter, real-world data are commonly dirty [140], and therefore the approach used to generate the BKVs must be able to deal with data that contain errors and variations and still achieve the aims of the indexing step, namely to put similar records into the same block, or closely together in the sorted databases.

As the example given under the attribute quality issue described above illustrated, an error in an attribute value used as a blocking key will lead to a record potentially being inserted into a different block. A commonly used way to overcome this problem is to define several different blocking keys, ideally based on different attributes, rather than having one blocking key only. The union of all candidate record pairs generated by each of the blocking key definitions is used in the comparison step to perform the detailed comparisons between records. Figure 4.1 illustrates a set of four example records and three blocking key definitions applied on them.

An alternative approach is to run the indexing step several times using different blocking key definitions (sometimes called 'blocking passes' [287]), and to compare and classify the generated candidate record pairs, with pairs classified as matches being removed from the input databases (or their records flagged as being matched).

RecID	GivenName	Surname	Postcode	Suburb
r1	peter	christen	2010	north sydney
r2	paul	smith	2600	canberra
r3	pedro	kristen	2000	sydeny
r4	pablo	smyth	2700	canberra sth

RecID	PC+Sndx(GiN)	Fi2D(PC)+DMe(SurN)	La2D(PC)+Sndx(SubN)
r1	2010-p360	**20-krst**	10-n632
r2	2600-p400	26-sm0	**00-c516**
r3	2000-p360	**20-krst**	00-s530
r4	2700-p140	27-sm0	**00-c516**

Fig. 4.1 Example records in the upper table and their blocking key values (BKVs) in the lower table, adapted from [64]. The first blocking key definition concatenates postcode (PC) values with Soundex (Sndx) encoded given name (GiN) values, the second blocking key definition concatenates the first two digits (Fi2D) of postcode values with Double-Metaphone (DMe) encoded surname (SurN) values and the third concatenates the last two digits (La2D) of postcodes with Soundex encoded suburb name (SubN) values. The hyphens ('-') in the BKVs are only shown for illustration, in real-world applications they would not be inserted. The two bold highlighted pairs show that records r1 and r3 would be inserted into the block with key '20-krst', and records r2 and r4 into the block with key '00-c516'

This approach also allows that different comparison and classification functions can be used in the different blocking passes.

The objective of using several blocking key definitions is that in at least one of them no errors or variations occur in the BKV, and thus a record is inserted into the 'correct' block and compared with those records that likely match with it. A second advantage of this approach is also that more selective blocking key definitions can be used, as was discussed above in the description of the trade-off between block numbers and their sizes. These will result in smaller blocks that are more specific and group more similar records together. And because several blocking key definitions are used, the likelihood increases that a pair of records that refers to a match has at least one BKV in common, as illustrated in Fig. 4.1.

All the indexing techniques that will be discussed in the remainder of this chapter do require the definition of a blocking or sorting key. An optimal definition of a blocking key would result in (1) all true matches being included in the candidate record pairs generated while (2) the total number of candidate record pairs generated is kept as small as possible. Blocking keys should generally be defined by keeping in mind the indexing technique that will be employed.

While traditionally blocking keys were defined manually by somebody who ideally has expertise in both data matching techniques and the domain of the data that are to be matched or deduplicated (especially the quality and characteristics of the data), several techniques have recently been proposed that allow the learning of optimal blocking keys from training data [34, 188]. These techniques are based on supervised machine learning algorithms and require training data in the form of pairs of records that are known to refer to true matches or true non-matches. Having such training data that need to be of high quality and diverse enough to cover as many true matches as possible, are however hard to get in many practical data matching applications. Therefore, the manual definition of blocking keys is still a widespread undertaking. These learning-based techniques will be described in more detail in Sect. 4.12.

4.3 (Phonetic) Encoding Functions

Functions to (phonetically) encode attribute values before they are used as blocking or sorting key values are commonly used in the indexing step of data matching and deduplication to bring similar sounding string values, that are often assumed to refer to names, into the same blocks. The records that contain these similar sounding names will then be compared in detail in the comparison step.

Phonetic encoding functions can however also be used in the comparison step to calculate the similarity between similar sounding string values, as will be discussed further in Sect. 5.2.

The common idea behind all phonetic encoding functions is that they attempt to convert a string, commonly assumed to refer to a name, into a code according to how a name is pronounced, i.e. how a name would be spoken [57]. This encoding process is often language dependent. Most techniques that have been developed (including all techniques presented in this chapter), are based on the assumption that names originate from the English language. Some of these techniques have been adapted for other languages [175, 238]. Other techniques, such as Double-Metaphone discussed below, can generate two encodings of a single name, depending upon whether there are variations in the spelling of a name.

4.3.1 Soundex

The Soundex [145, 175, 302] algorithm is one of the oldest approaches. It was developed and patented by Russell and Odell in 1918 [201], and is one of the best known and most widely used phonetic encoding algorithm. Based on American-English language pronunciation, it encodes name strings by keeping the first letter in a string and converting the remaining characters of the string into numbers according to the transformation table given in Fig. 4.2.

After transforming a string into digits, all zeros (which correspond to vowels and 'h', 'w' and 'y') are removed from the encoded string, and all repetitions of the same number are also removed. For example, an initial transformed encoding of 'p0330111' is converted into 'p31', and the initial encoding 's550144042' is converted into 's5142'. If an encoding contains less than three digits (as the first example), then the code is extended with zeros to a total length of three digits (so 'p31' becomes 'p310'), while codes that contain more than three digits are truncated to three digits only (thus the encoding 's5142' becomes 's514').

The advantages of Soundex are its simplicity and computational efficiency. A first major drawback of Soundex is that a difference in the first letter of two name strings

Table 4.1 Example name strings and their phonetic encodings. Variations of the same name are grouped together

String	Soundex	Phonex	Phonix	NYSIIS	Double Metaphone	Fuzzy Soundex
peter	p360	b360	p300	pata	ptr	p360
pete	p300	b300	p300	pat	pt	p300
pedro	p360	b360	p360	padr	ptr	p360
stephen	s315	s315	s375	staf	stfn	s315
steve	s310	s310	s370	staf	stf	s310
smith	s530	s530	s530	snat	sm0, xmt	s530
smythe	s530	s530	s530	snat	sm0, xmt	s530
gail	g400	g400	g400	gal	kl	g400
gayle	g400	g400	g400	gal	kl	g400
christine	c623	c623	k683	chra	krst	k693
christina	c623	c623	k683	chra	krst	k693
kristina	k623	c623	k683	cras	krst	k693

results in different Soundex codes for these names, as can be seen in Table 4.1 for the name strings 'christina' and 'kristina' with their corresponding Soundex codes 'c623' and 'k623', respectively. A second drawback is that Soundex codes are mostly representing the beginning of name strings, and differences that appear towards the end of two names are often not represented properly because they are pruned away if the codes are too long. A commonly applied solution to both these drawbacks is to not only generate the Soundex encodings of name strings, but also the encodings of the reversed name strings. A name pair is then seen to be similar if either of the two calculated encodings are the same.

4.3.2 Phonex

This encoding algorithm is a variation of the original Soundex approach [175]. It aims to improve the quality of the calculated encodings through a pre-processing step where name strings are modified according to their English pronunciation before Soundex-like encodings are generated. The following modifications are applied to a name string in the pre-processing step:

- All 's' characters at the end are removed.
- A 'kn' character sequence at the beginning is replaced with a single 'n' character.
- A 'ph' character sequence at the beginning is replaced with a single 'f' character.
- A 'wr' character sequence at the beginning is replaced with a single 'r' character.
- An 'h' character at the beginning of a name string is removed.
- If the first character is a vowel (including 'y') then it is replaced with an 'a'.
- If the first character is a 'p' then it is replaced by a 'b'.
- If the first character is a 'v' then it is replaced by an 'f'.
- If the first character is a 'k' or a 'q' then it is replaced by a 'c'.

- If the first character is a 'j' then it is replaced by a 'g'.
- If the first character is a 'z' then it is replaced by an 's'.

After this initial pre-processing step, the processed name string is encoded similar as with Soundex into a code made of the initial character followed by three digits. The transformation from letters into digits is somewhat different from the original Soundex transformation, because several transformation rules are taken into account. Similar to the initial pre-processing done, these rules take character sequences into account when converting letters into digits [175].

4.3.3 Phonix

The Phonix algorithm extends the idea of Phonex pre-processing of name strings even further by applying more than a hundred transformation rules. These rules are applied not only on a single character, but also on sequences of several characters. While most rules are applied anywhere in a name string, 19 rules are only applied if the character(s) appear(s) at the beginning of a string, 12 rules are applied only to the middle of a string, and 28 rules only to the end of a string.

Similar to Soundex and Phonex, the transformed name string is encoded into a code consisting of a starting letter followed by three digits (again removing zeros and duplicate numbers). The transformation table is different from the one used in Soundex, as can be seen from Fig. 4.3.

The larger number of transformation rules means that the Phonix algorithm is more complex and thus slower than the Soundex and Phonex algorithms. An experimental evaluation has shown that Phonix is around ten times slower than Soundex on different data sets that contained several thousand name strings each [57].

4.3.4 NYSIIS

The *New York State Identification and Intelligence System* (NYSIIS) phonetic encoding algorithm departs from the one-letter three-digit code and only returns an encoding made of letters [40]. Similar to Phonex and Phonix, it applies various rules to the input name string. These rules are:

- Transform various beginnings of the name string: 'mac' becomes 'mcc', 'kn' becomes 'n', 'k' is replaced with 'c', 'ph' and 'pf' are replaced with 'ff', and 'sch' with 'sss'.
- Transform various endings of the name string: 'ee' and 'ie' are replaced with 'y', while 'dt', 'rt', 'rd', 'nt' and 'nd' are all replaced with 'd' only.
- The first letter of the transformed name string now becomes the first letter of the NYSIIS encoding.
- The remaining letters of the transformed name string are further transformed using one of the following rules, applied starting from the beginning of the string:

Fig. 4.2 Soundex encoding
transformation table

a, e, h, i, o, u, w, y	$\rightarrow 0$
b, f, p, v	$\rightarrow 1$
c, g, j, k, q, s, x, z	$\rightarrow 2$
d, t	$\rightarrow 3$
l	$\rightarrow 4$
m, n	$\rightarrow 5$
r	$\rightarrow 6$

Fig. 4.3 Phonix encoding
transformation table

a, e, h, i, o, u, w, y	$\rightarrow 0$
b, p	$\rightarrow 1$
c, g, j, k, q	$\rightarrow 2$
d, t	$\rightarrow 3$
l	$\rightarrow 4$
m, n	$\rightarrow 5$
r	$\rightarrow 6$
f, v	$\rightarrow 7$
s, x, z	$\rightarrow 8$

1. 'ev' is replaced by 'af'.
2. 'e', 'i', 'o' and 'u' are replaced with 'a'.
3. 'q' is replaced by 'g', 'z' is replaced by 's' and 'm' is replaced by 'n'.
4. 'kn' is replaced by 'n' and 'k' by 'c'.
5. 'sch' is replaced by 'sss' and 'ph' by 'ff'.
6. If the letter before or after an 'h' is not a vowel then the 'h' is replaced by the letter before it.
7. If the letter before a 'w' is a vowel then the 'w' is replaced with 'a'.
8. Only add the current processed letter to the NYSIIS encoding if it is different from the previous letter in the encoding.

- Several rules are then applied to the end of the encoding:

1. If the last letter in the encoding is an 'a' or 's' then remove it.
2. If the encoding ends with 'ay' replace it by 'y'.

- Finally, if the length of the encoding is longer than six letters then truncate it to the first six letters only.

Besides Soundex, the NYSIIS algorithm is the second most popular phonetic encoding algorithm employed for data matching and deduplication, as well as other applications that require the grouping of similar sounding names strings.

4.3.5 Oxford Name Compression Algorithm

The Oxford Name Compression Algorithm (ONCA) combines the NYSIIS and Soundex algorithms [118, 119]. The ONCA has been used in the Oxford Record Linkage System. In a first step, name strings are processed using a version of the NYSIIS algorithm that was adapted for Anglo-Saxon and European names. The

Table 4.2 Examples of the ONCA phonetic encoding algorithm, adapted from [119]

Original string	NYSIIS encoding	ONCA encoding
andersen, anderson	andar	a536
brian, brown, brun	bran	b650
capp, cope, copp, kipp	cap	c100
dane, dean, dent, dionne	dan	d500
smith, schmit, schmidt	snat	s530
truman, trueman	tranan	t655

resulting phonetic codes are further processed by applying the standard Soundex algorithm on them. Table 4.2 shows several examples of the ONCA approach.

4.3.6 Double-Metaphone

A major drawback of the four phonetic encoding algorithms discussed so far is that they are specifically aimed at English names, and are therefore not suitable for databases that contain names from different languages. Many countries have an increasingly multi-cultural population, and therefore non-English names appear more frequently in many databases that contain detailed information about people. It is therefore important that a phonetic encoding algorithm can accommodate non-English names.

The Double-Metaphone algorithm attempts to accomplish this by better accounting for European and Asian names [211]. Similar to the Phonix and NYSIIS algorithms, a large number of transformation rules are applied to a name string. These rules take the position within a name string into account, and some rules also consider the previous and following letters. In line with the NYSIIS encoding, Double-Metaphone returns an encoding made of letters only. Different from the NYSIIS algorithm, however, is that for certain name strings not only one phonetic encoding is calculated but two. These two codes are based on the application of different phonetic transformation rules. For example, the Polish name 'kuczewski' will be encoded as 'kssk' and 'kxfsk', accounting for different spelling variations of this name. In general, Double-Metaphone seems to be generating encodings that are closer to the correct pronunciation of names than NYSIIS.

4.3.7 Fuzzy Soundex

This algorithm combines a q-gram based pre-processing step with a Soundex like transformation table [145]. Q-grams are substrings of length q. In the fuzzy Soundex algorithm, q-grams of length 2 (bigrams) and 3 (trigrams) are applied similar to the substitutions applied in Phonix, NYSIIS or Double-Metaphone. Some of these substitutions are only applied at the beginning of a name string, while others are

Fig. 4.4 Fuzzy Soundex
encoding transformation table

a, e, h, i, o, u, w, y	$\rightarrow 0$
b, f, p, v	$\rightarrow 1$
d, t	$\rightarrow 3$
l	$\rightarrow 4$
m, n	$\rightarrow 5$
r	$\rightarrow 6$
g, j, k, q, x	$\rightarrow 7$
c, s, z	$\rightarrow 9$

applied anywhere. The pre-processed name string is then converted into a one-letter three-digit encoding using the transformation table shown in Fig. 4.4.

Fuzzy Soundex was developed within the information retrieval community with the aim to improve the quality of Soundex-based retrieval [145]. Combined with a q-gram based pattern matching algorithm, it achieved better retrieval results on a database of over 30,000 names than the basic Soundex algorithm [210].

4.3.8 Other Encoding Functions

The encoding functions discussed so far are all aimed at the phonetic encoding of strings that are assumed to be names, such as personal or address names. As different types of data are commonly being used in data matching, some forms of encoding functions (possibly not phonetic) need to be available for data that do not correspond to names.

Recall that the objective of an encoding function is to bring 'similar' values together. For data that are not name strings, a 'binning' type of encoding function can be employed. Commonly used to smooth noisy data [135], binning puts numerical values that are similar to each other into the same bin, an approach related to blocking.

For example, numerical age values (as years) can be binned by having one bin (block) per age decade, thus inserting all records that have an age value from 0–9 into one bin, those with an age value of 10–19 into a second bin and so on. For postcode values, as illustrated in Fig. 4.1, blocking can be achieved by only taking a subset of the available digits, such as only the first two or only the last two out of four postcode digits. This leads to a maximum of 100 bins. During the indexing process, all records that have the same first two (or last two) digits in common are inserted into the same block. If these blocks become too big, then taking the first or last three digits (leading to maximum 1000 bins and blocks) is an alternative.

For date values, depending upon the distribution and spread of date values in a database (i.e. the difference between the first and last date), either year values only, or month and year values combined can be used as the encoding function, resulting for example in blocking key values such as 'jan2011', 'feb2011' and so on.

Database A

RecID	GivenName	Surname	Sndx(GiN)	Sndx(SurN)
a1	peter	myler	p360	**m460**
a2	pedro	smith	p360	s530
a3	steve	peters	**s315**	p362
a4	gail	smythe	g400	s530
a5	christine	miller	c623	**m460**

Database B

RecID	GivenName	Surname	Sndx(GiN)	Sndx(SurN)
b1	kristina	miller	k623	**m460**
b2	stephen	peter	**s315**	p360
b3	kylie	smith	k400	s530
b4	pete	myler	p300	**m460**
b5	kellie	roberts	k400	r163

Candidate record pairs from GivenName

BKV	Candidate record pairs
s315	(a3,b3)

Candidate record pairs from Surname

BKV	Candidate record pairs
m460	(a1,b1), (a1,b4), (a5,b1), (a5,b4)
s530	(a2,b3), (a4,b3)

Blocks A: GivenName

Blocks A: Surname

Blocks B: GivenName

Blocks B: Surname

Fig. 4.5 Two example databases with their blocking key values (BKVs) based on the Soundex (Sndx) encoded given name (GiN) and surname (SurN) values, the blocks generated using the standard blocking approach, and the resulting candidate record pairs. The BKVs that are generated from both databases are highlighted in boldface

4.4 Standard Blocking

This traditional indexing approach has been used in data matching and deduplication for several decades [108]. The uniqueness of this approach is that the identifier of each record is inserted into one block only. All other indexing techniques presented in this chapter potentially insert a single record into several blocks.

Assuming a single blocking key has been defined, one blocking key value (BKV) will be generated for each record in the input database(s). This BKV determines into which block a record is inserted. All records that have the same BKV are inserted into the same block. For the matching of two databases, pairs of candidate records are generated from all records that have the same BKV across both databases. If a BKV occurs in records from one database only, then no record pairs will be formed from this block, as there are no records in the corresponding block in the other database.

For a deduplication, pairs of candidate records are generated from all unique pairs of record identifiers within a block. Because the comparison of two records is symmetric, each unique record pair only needs to be compared once. For example, with a block that contains the three record identifiers 'r1', 'r2' and 'r3', the generated record pairs for a deduplication would be (r1,r2), (r1,r3) and (r2,r3), but not (r2,r1), (r3,r1) or (r3,r2).

An efficient way to implement standard blocking is to build an inverted index data structure [288, 303], where each BKV becomes the key of an index list, and the identifiers of all records that are in the same block are inserted into the same inverted index list. Figure 4.5 illustrates such inverted index lists for two small example databases.

As discussed in Sect. 4.2 before, several blocking keys are generally defined (often on different attributes), and for each a separate index data structure is built. Candidate record pairs are then generated independently when blocks are processed from each index data structure. However, even if a candidate record pair is generated several times, the corresponding record pair will only be compared once in the comparison step.

The number of candidate record pairs that are generated with standard blocking depends upon the frequency distribution of the BKVs [64]. The most frequent BKVs will generate the most candidate record pairs. For a simplified estimate, a uniform frequency distribution of BKVs can be assumed. If the number of records in the two databases to be matched is denoted with m and n, respectively, and the number of BKVs in common with b, then each block contains m/b or n/b records, respectively. The total number of candidate record pairs generated, c, then is

$$c = b \left(\frac{m}{b} \cdot \frac{n}{b} \right) = \frac{m \cdot n}{b}. \tag{4.1}$$

For the deduplication of a single database that contains n records (and b BKVs), the number of candidate record pairs generated, c, is

$$c = b \left(\frac{n}{b} \cdot \frac{n-1}{b} \right) /2 = \frac{n \cdot (n-1)}{2b}. \tag{4.2}$$

A more detailed complexity analysis for other frequency distributions of the BKVs can be found in the recent survey by Christen [64].

4.5 Sorted Neighbourhood Approach

The first alternative indexing technique to standard blocking was developed in the mid-1990s by Hernandez and Stolfo [140, 141]. Rather than generating blocks according to the BKVs, this approach sorts the databases to be matched according to a 'sorting key' (which is generated in a similar way as a blocking key). A sliding window of fixed size w (with $w > 1$) is then moved over the sorted databases, and candidate record pairs are generated from the records that are in the window in any given step.

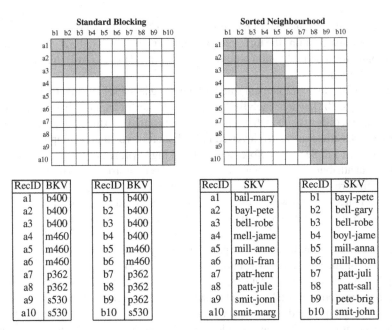

Fig. 4.6 Comparison of the candidate record pairs that are generated (shaded squares) with standard blocking and the sorted neighbourhood approach. For standard blocking it is assumed the records are sorted according to the BKVs to better illustrate the blocks generated. The window size for the sorted neighbourhood approach is set to $w = 3$. Note that a different blocking key and sorting key definition is used for the two approaches. For standard blocking, Soundex codes on surnames are used, while for the sorted neighbourhood approach the sorting key values (SKVs) are the first four letters of surnames concatenated with the first four letters of given names. The hyphen ('$-$') is only used for illustration, in a practical implementation values would be concatenated directly

Figure 4.6 illustrates both the standard blocking and the sorted neighbourhood approach. The first position of the sliding window, with $w = 3$, contains the records 'a1' to 'a3' and 'b1' to 'b3', and therefore the nine candidate record pairs (a1,b1), (a1,b2), (a1,b3), (a2,b1), (a2,b2), (a2,b3), (a3,b1), (a3,b2) and (a3,b3) will be generated. In window position 2 (which contains records 'a2' to 'a4' and 'b2' to 'b4'), the four candidate record pairs (a2,b2), (a2,b3), (a3,b2) and (a3,b3) will again be generated. However, each unique candidate record pair will only be compared once in the comparison step. The new candidate record pairs generated in the second window position will be (a2,b4), (a3,b4), (a4,b2), (a4,b3) and (a4,b4).

In the approach originally proposed by Hernandez and Stolfo [140], both databases are merged before they are sorted according to their sorting key values (SKVs), and then the sliding window is moved over the combined SKVs. In this way, similar records from both databases are moved closer together in the hope that they are included in the same window.

The criteria used to define a good sorting key is different from the criteria used to define a good blocking key. While a blocking key has to balance the quality of the candidate record pairs generated (how many true matches are included) with the size of the blocks generated (and thus the number of resulting candidate record pairs), the definition of a sorting key needs to be more specific and fine-grained, such that sorting the databases brings similar records closer together.

An important consideration when defining a sorting key is that the sorting of the databases is very sensitive to the beginning of the SKVs, especially the first character. For example, the two similar given name values 'christine' and 'kristina' will not be close to each other if the sorting is based on given names. Similar to the multiple blocking key definitions commonly used for standard blocking, it is of advantage to run several iterations of the sorted neighbourhood approach using different sorting key definitions [140].

The number of candidate record pairs that will be generated with the sorted neighbourhood approach does not depend upon the frequency distribution of the SKVs. Assuming both databases to be matched contain n records, and the window size is set to w, then there will be $(n - w + 1)$ window positions. In the first position, the number of candidate record pairs that are generated is $w \cdot w = w^2$, while in each of the following positions $w + (w - 1) = 2w - 1$ new unique candidate pairs are added to the set of candidate pairs (all other pairs in a certain window position have already been generated in previous window positions). Overall, the total number of unique candidate records pairs will be

$$c = w^2 + (n - w)(2w - 1) = 2nw - 2w^2 - n. \tag{4.3}$$

For large databases, it holds that $w \ll n$, and therefore this process has a computation complexity of $O(n)$, i.e. it is linear in the size of the databases to be matched. The same holds for the deduplication of a single database. However, the sorting of the SKVs must also be considered. This can be accomplished in $O(n \log n)$ steps.

An alternative of how the sorted neighbourhood approach can be implemented was recently proposed [64]. Depending upon how the SKVs are defined, there might be many SKVs that are the same, and therefore the sliding window will not cover all records with the same SKV. For example, in a large database there might be many records with surname 'smith' and given name 'john', resulting in many records with SKV 'smith-john' (similar to the example shown in Fig. 4.6). The proposed alternative is to generate an inverted index data structure (as can be done with standard blocking), where the index keys are the unique SKV values. The index keys are then sorted, and a sliding window is moved over the index key values rather than the SKVs directly. Each unique SKV only appears once in the sorted index key values, while with the original sorted neighbourhood approach an SKV occurred as many times as a record in the database(s) contained the SKV. At any window position, the union of the record identifiers from all inverted index lists in the current window will be used to generate the candidate record pairs for that window position. Because the inverted index lists can have different lengths, the number of candidate record pairs that are generated depends upon the distribution of the attribute values used in the SKVs rather than the window size only. Experimental results have shown that this alternative approach can lead to more true matches being included into the set of candidate record pairs at the cost of a larger number of generated candidate pairs [64].

A major drawback of the original sorted neighbourhood approach is that the fixed window size can result in missed true matches if similar records are not close enough in the sorted SKV array to be in the same window. This drawback has been addressed by a recent approach that dynamically changes the window size w according to the values of the SKVs [292]. The window size is increased as long as SKVs in the sorted array are similar to each other according to an approximate string comparison function (as will be discussed in Chap. 5). A window will cover a sequence of SKVs (and their records) that have a similarity between each other above a certain similarity threshold. A new window will be started at a 'boundary pair' where two consecutive SKVs have an approximate string similarity below a certain similarity threshold.

Other recent work has generalised the standard blocking and sorted neighbourhood approach and shown that they can be two ends of the same approach [94]. Standard blocking can be seen as the sorted neighbourhood approach where the window moves w positions forward rather than only 1, leading to non-overlapping blocks. A novel indexing technique based on this approach has been proposed that allows the specification of a desired overlap as well as window size [94]. An experimental evaluation of this technique showed that the sorted neighbourhood approach outperformed standard blocking, especially when blocks were set to a small size [94].

4.6 Q-Gram Based Indexing

For data that are dirty and contain large amounts of errors and variations, both standard blocking and the sorted neighbourhood approach might not be able to insert records into the same blocks, for example if the beginning of a sorting key value is different for two name variations. Q-gram based indexing aims to overcome this drawback by generating variations of each BKV, and to use these variations as the actual index keys for a standard blocking-based indexing approach. Each record is inserted into several blocks according to the variations generated from its BKV [20].

Q-gram based indexing takes each blocking (or sorting) key value and converts it into a list of q-grams. A q-gram (also known as n-gram [172]) is a substring of length q characters. Common choices for q are $q = 2$ (called *bigrams* or *digrams* [162]) or $q = 3$ (called *trigrams* [258]). A string s that is $c = |s|$ characters long contains $k = c - q + 1$ q-grams. The list of q-grams of a string s is generated using a sliding window approach that extracts q characters from s at any position from 1 to k of the string. For example, the bigram list that is generated from the string 'christen' is ['ch', 'hr', 'ri', 'is', 'st', 'te', 'en'].

To create variations of a BKV, sub-lists of the q-gram list are generated in a recursive approach, as illustrated in Fig. 4.7. If the original q-gram list contains k q-grams, then in the first step k sub-lists of length $k - 1$ q-grams are generated. In each of these sub-lists, one q-gram is removed. The process is then applied to each

RecID	BKVs (Surname)	Bigram sub-lists	Index key values
r1	miller	[mi,il,ll,le,er], [il,ll,le,er], [mi,ll,le,er], [mi,il,le,er], [mi,il,ll,er], [mi,il,ll,le], [ll,le,er], [il,le,er], [il,ll,er], [il,ll,le], [mi,le,er], [mi,ll,er], [mi,ll,le], [mi,il,er], [mi,il,le], [mi,il,ll]	'miilller', 'illlleer', 'millleer', 'miilleer', 'miiller', 'miilllle', '**llleer**', 'illeer', 'iller', 'miller', 'illlle', 'mileer', 'millle', 'miiler', 'miille', 'miilll'
r2	muller	[mu,ul,ll,le,er], [ul,ll,le,er], [mu,ll,le,er], [mu,ul,le,er], [mu,ul,ll,er], [mu,ul,ll,le], [ll,le,er], [ul,le,er], [ul,ll,er], [ul,ll,le], [mu,le,er], [mu,ll,er], [mu,ll,le], [mu,ul,er], [mu,ul,le], [mu,ul,ll]	'muullleer', 'ulllleer', 'mullleer', 'muulleer', 'muuller', 'muulllle', '**llleer**', 'ulleer', 'uller', 'muller', 'ulllle', 'muleer', 'mullle', 'muuler', 'muulle', 'muulll'

Fig. 4.7 Q-gram based indexing with two surname values used as BKVs. Q-grams of length $q = 2$ (bigrams) are used, and the minimum threshold is set to $t = 0.75$. Duplicate q-gram sub-lists are removed. The index key value that is generated for both records is highlighted in boldface

of these sub-lists in a recursive manner. A minimum threshold t ($t < 1$) is set by a user to decide the minimum relative length, l, of the shortest q-gram sub-lists that are to be generated. For a BKV that contains k q-grams and with a threshold set to t, all sub-lists down to a length

$$l = max(1, \lfloor k \cdot t \rfloor) \tag{4.4}$$

are generated, with $\lfloor \ldots \rfloor$ denoting the rounding to the next lower integer value. All sub-lists are then converted back into strings and used as the actual index keys in an inverted index data structure, as described for standard blocking and illustrated in Fig. 4.5. Each record is inserted into several inverted index lists, according to how many index keys have been generated from its BKV [64].

As shown in Fig. 4.7, a major drawback of this indexing approach is that (even with short BKVs) a large number of sub-lists (and thus index keys) are generated, and each record is likely inserted into many blocks. The recursive generation of sub-lists is a computationally expensive procedure, especially for long BKVs and low threshold values.

The advantage of q-gram based indexing is that it can overcome errors and variations in the BKVs, and therefore records that refer to true matches are more likely inserted into the same index list, even if their BKVs are different from each other. This leads to more true matching records being compared and thus an improved matching quality.

A recent theoretical and empirical evaluation of q-gram based indexing has illustrated that the drawback of having a high computation complexity outweighs the advantage of being able to match data that are dirty [64]. Q-gram based indexing is not suitable for the matching or deduplication of large databases because generating the many q-gram sub-lists takes a prohibitive amount of time.

An approximate string join technique within a database framework that is related to q-gram based indexing was proposed by Gravano et al. [127]. It uses q-grams to reduce the computation complexity of the naive pair-wise approach of comparing all possible record pairs when joining two database tables. This technique augments a database with a table that for each record contains tuples that are made of the identifier of the record, the q-grams extracted from the record's attribute value and the positions of these q-grams within the attribute value (positional q-grams will be discussed further in Sect. 5.4). Using this q-gram table, an SQL query is used to only compare candidate record pairs that fulfil three filtering criteria. These criteria limit the number of pairs that have to be compared using an expensive user-defined function (UDF) such as edit distance (which will be discussed in Sect. 5.3). The first criteria, count filtering, selects pairs that have a certain minimum number of q-grams in common. The second criteria, position filtering, removes pairs where the common q-grams are at positions too far apart. Finally, the third criteria, length filtering, removes pairs that have a length difference of their corresponding strings above a certain threshold. An experimental evaluation by the authors using a commercial database showed that this proposed approach results in a very effective approximate string join implemented completely within a relational database [127].

4.7 Suffix-Array Based Indexing

This indexing technique is related to q-gram based indexing. It also aims to overcome errors and variations in the BKVs by generating suffix substrings (called suffixes) of the BKVs. The suffixes of a string are all its substrings with one or more characters at the beginning removed. For example, the suffixes of the string 'peter' are 'eter', 'ter', 'er' and 'r'. Each unique suffix string becomes the key of an index block, and all records that contain this suffix string are inserted into this block [7]. Similar to q-gram based indexing, the identifier of a record will likely be inserted into several inverted index lists. Figure 4.8 illustrates this approach on four example BKVs.

When applied for indexing, the shorter a suffix string is the more BKVs (and thus more records) will contain this suffix. This will result in very large blocks. For example, the identifiers of all records that contain a BKV that ends with the suffix 'r' will be inserted into the block with index key 'r'. To prevent such large blocks to occur, suffix-array based indexing has two parameters that influence the size of the index blocks that are generated.

- The first parameter is the minimum suffix length, l_{min}. This parameter sets the minimum length of suffix strings that are generated. With $l_{min} = 4$, for example, for the string 'christen' the following suffixes will be generated: 'hristen', 'risten', 'isten' and 'sten'. The identifier of each record that has the BKV 'christen' will therefore be inserted into 5 blocks (the value 'christen' will also be used as a key of

RecID	BKVs (GivenName)	Suffixes
r1	katherina	katherina, atherina, therina, herina, erina, rina
r2	catrina	catrina, atrina, trina, rina
r3	catherine	catherine, atherine, therine, herine, erine, rine
r4	catherina	catherina, atherina, therina, herina, erina, rina
r5	katrina	katrina, atrina, trina, rina

Suffix	RecID
atherina	r1, r4
atherine	r3
atrina	r2, r5
catherina	r4
catherine	r3
catrina	r2
erina	r1, r4
erine	r3
herina	r1, r4
herine	r3
katherina	r1
katrina	r5
~~rina~~	~~r1, r2, r4, r5~~
rine	r3
therina	r1, r4
therine	r3
trina	r2, r5

Fig. 4.8 Suffix-array based indexing example of five given name values, adapted from [64]. The minimum suffix length is set to $l_{min} = 4$ and the maximum block size to $b_{max} = 3$. As a result, the block for suffix 'rina' is too large (because four records contain this suffix value) and it is deleted before the candidate record pairs are generated

the inverted index). For BKVs that are shorter than l_{min} characters (such as 'tan'), only their actual value will be used as index key.

A BKV that is c characters long will result in $k = (c - l_{min} + 1)$ suffix strings, and therefore a record that has a BKV of length c will be inserted into k inverted index lists and thus blocks. The longer a suffix string is, the less likely it will occur frequently in all the BKVs in a database. For example, there will be more records in a database that contain the suffix value 'tina' in their BKVs compared to 'ttina' (from given name 'bettina'), because 'christina', 'kristina', 'martina' and 'santina' also contain the suffix 'tina'.

- To limit the size of the generated blocks, the second parameter used in suffix-array based indexing is the maximum block size that is allowed, b_{max}. Once the BKVs and their suffix strings have been generated for all records in a database and the record identifiers have been inserted into the suffix array index, then all index blocks that contain more than b_{max} record identifiers will be deleted. These large blocks were likely generated by short suffix strings that appear in many BKVs. By removing these large blocks, the number of candidate record pairs that are generated is limited. The idea of this pruning step is that each record identifier is likely being inserted into several index blocks. Even after the large blocks have been removed, a record identifier is still kept in one or more of the smaller blocks. In case of where a large block is deleted, there is, however, a chance that such a block contained record identifiers where their BKV was only inserted into this large block (because the BKV for these records was of length l_{min}). As a result, these records would not be part of any candidate record pair, and thus would not

be compared with other records. Therefore, in such a case a large block should not
be deleted completely, but its size (number of record identifiers in it) be reduced
by only removing record identifiers that have the longest original BKV (i.e. that
have also been inserted in several other blocks).

Suffix-array based indexing has successfully been applied to the deduplication of
both English and Japanese bibliographic databases, where suffix arrays were created
based on English names and Japanese characters, respectively [7].

The number of candidate record pairs, c, that will be generated with this indexing
technique can be estimated assuming that all blocks that are generated contain the
maximum allowed number of record identifiers, b_{max}. If b such blocks are generated,
then in total

$$c = b \cdot b_{max}^2 \qquad (4.5)$$

candidate record pairs are generated for the matching of two databases, and

$$c = b \cdot (b_{max}(b_{max} - 1)) / 2 \qquad (4.6)$$

for the deduplication of one database [64]. In practice this will likely be an upper
bound, because not all blocks will reach the maximum allowed size b_{max}. Compared
to standard blocking, the number of blocks b will also be much larger with suffix-array
based indexing, because each BKV used in standard blocking will likely generate
several different suffix values.

As Fig. 4.8 illustrates, the suffix-array based indexing technique has the drawback that variations or errors at the end of BKVs will lead to record identifiers being
inserted into different blocks, potentially resulting in missed true matches. This is
especially of concern as empirical studies have shown that more data entry errors
appear at the end of strings compared to their middle or beginning [214]. A modification to the suffix generation process can help overcome this drawback. Rather
than only generating the suffix strings of a BKV, all substrings down to the minimum
length l_{min} can be generated in a sliding window fashion (similar to the generation of q-grams). For example, for the string 'christen' and $l_{min} = 5$, this approach
would generate the substrings: 'christen' (length 8); 'christe' and 'hristen' (length
7); 'christ', 'hriste', and 'risten' (length 6); and 'chris', 'hrist', 'riste', and 'isten'
(length 5) [64]. This approach can help to overcome errors both at the beginning
and end of BKVs, but the computational cost (similar to q-gram based indexing)
increases significantly because a larger number of substrings are generated, and a
record identifier is inserted into a larger number of blocks.

In an approach that is similar to the adaptive sorted neighbourhood technique
[292] discussed in Sect. 4.5, a recently developed improvement to suffix-array based
indexing is to merge blocks if their suffix values are similar to each other [265, 266].
This merging is based on an approximate string similarity function that calculates
the similarity between consecutive strings in the sorted array of suffix values. If the
similarity of a string pair is above a certain threshold, then their corresponding lists

of record identifiers are merged to form a new combined larger block. As a result, suffix-array based indexing becomes more robust [265, 266].

As an example, assume the suffix array from the right-hand side of Fig. 4.8 has been generated. If the edit distance string comparison function (which returns a normalised similarity value between 0.0 and 1.0, as will be discussed in detail in Sect. 5.3) with a minimum similarity threshold set to $t = 0.85$ is used, three pairs of neighbouring suffix strings have a similarity above this threshold. The string pair 'atherina' and 'atherine' has a similarity $s = 0.875$, and thus their record identifier lists are merged into the list [r1,r3,r4]. The pair 'catherina' and 'catherine' has a similarity $s = 0.889$, and their lists are merged into [r3,r4]. Finally, the string pair 'therina' and 'therine' has a similarity $s = 0.857$ and their merged list is [r1,r3,r4].

A detailed theoretical analysis of this robust suffix-array based indexing technique has been presented by the developers [265]. A recent extension of this technique is to employ Bloom filters to improve the computational performance and reduce the amount of main memory that is required [266]. An experimental evaluation showed that with Bloom filters the number of database accesses needed during the generation of the index data structure can be reduced by up to 70 %.

4.8 Canopy Clustering

The indexing step can be seen as a clustering of the records in the databases to be matched or deduplicated in such a way that records that are similar to each other are inserted into the same cluster. Many clustering algorithms have high computation complexity [135]. Indexing, however, should be computationally cheap, and it must be feasible to generate the candidate record pairs in a fast and scalable manner. The canopy clustering approach achieves this goal by efficiently calculating distances between the BKVs [85, 185], and inserting records into one or more overlapping clusters. Each cluster then becomes a block from which candidate record pairs are generated.

The similarities between BKVs are calculated using either the Jaccard or the TF-IDF/Cosine (Term-Frequency / Inverse Document Frequency [288]) similarity measures, using tokens generated from the BKVs. Both these measures are also used in approximate string comparison functions, as will be described in detail in Chap. 5. Jaccard similarity, $sim_{Jaccard}$, is a measure based on the number of tokens two BKVs, b_1 and b_2, have in common, normalised by the union of the tokens contained in the two BKVs. If the function $token(b)$ returns the set of tokens in the BKV b, then the Jaccard similarity is calculated as

$$sim_{Jaccard} = \frac{|token(b_1) \cap token(b_2)|}{|token(b_1) \cup token(b_2)|},$$ (4.7)

with $|\ldots|$ denoting the number of elements in a set [64]. When the TF-IDF/Cosine similarity, sim_{TFIDF}, is used instead of the Jaccard similarity, then TF-IDF weight-

RecID	BKVs (Surname)	Sorted bigram lists
r1	hanlan	[(an,2), (ha,1), (la,1), (nl,1)]
r2	gansan	[(an,2), (ga,1), (ns,1), (sa,1)]
r3	gargan	[(an,1), (ar,1), (ga,2), (rg,1)]

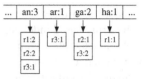

Fig. 4.9 Canopy clustering example with BKVs based on surnames. The tokens used are the bigrams extracted from these surname values. The sorted bigram lists include Document Frequency (DF) counts. The Term Frequency (TF) and DF counts in the inverted index data structure in the right-hand side are used to calculate weights in the TF-IDF/Cosine similarity measure, while the Jaccard similarity measure will be calculated based only on the bigrams in the inverted index lists. This figure is adapted from [64]

ing (described below) is taken into account when the similarity between two BKVs is calculated.

Depending upon the content of the BKVs, the tokens used in canopy clustering can either be words or q-grams (substrings of length q, as was discussed in Sect. 4.6). The canopy clustering indexing technique is based on an inverted index data structure where the index keys are the tokens extracted from the BKVs, and the index lists contain the identifiers of the records that contain this token in their BKVs. Figure 4.9 shows a small example of such a data structure. The Document Frequency (DF) is the count of how many times a token appears in a BKV in a certain record. For example, the token 'an' appears twice in the BKV of records 'r1' and 'r2' in the example. The DF can be calculated as the BKV for each record is being processed. The Term Frequency (TF), is the count of how many records in a database contain a certain token (the token 'an' for example appears in all three records in Fig. 4.9). Once a database has been loaded and the inverted index data structure for canopy clustering has been generated, the TF values can be converted into Inverse Document Frequencies (IDF) [288]. While there are several variations of how IDF can be calculated [135, 288], the basic approach is using the following equation:

$$idf(t) = \frac{n}{tf(t)}, \tag{4.8}$$

where n is the number of records in the database and $tf(t)$ is the number of records in the database that contains the term t [288].

After the inverted index data structure is built, the canopy clustering indexing algorithm can be started [64]. The algorithm iteratively generates overlapping clusters by repeating the following steps [85].

1. The identifiers of all records in the databases to be matched are inserted into a set, P.
2. One record identifier, r_c, is randomly selected from P. This record identifier will become the centroid of a new canopy cluster, $C_i = \{r_c\}$.
3. Using the tokens in the BKV of record r_c, either the Jaccard, $sim_{Jaccard}$, or TF-IDF/Cosine similarity, sim_{TFIDF}, is calculated with all records $r_x \in P$

that have at least one token in common with r_c. Using the inverted index data structure, this can be accomplished very efficiently.

4. All records r_x that have a Jaccard or TF-IDF/Cosine similarity value above a loose threshold, t_l, are inserted in the canopy cluster C_i (i.e. either $sim_{Jaccard}(r_c, r_x) \geq t_l$ or $sim_{TFIDF}(r_c, r_x) \geq t_l$).

5. All records $r_x \in C_i$ that have a similarity with r_c above a tight threshold, t_t, are removed from the set P of records (i.e. either $sim_{Jaccard}(r_c, r_x) \geq t_t$ or $sim_{TFIDF}(r_c, r_x) \geq t_t$). The cluster centroid, r_c, is also removed from P.

6. As long as the set of records is not empty, $P \neq \emptyset$, go back to step 2.

The loose threshold, t_l, needs to be smaller or equal to the tight threshold, t_t, i.e. $t_l \leq t_t$. If both thresholds are set to the same value ($t_l = t_t$), then the generated canopy clusters will not be overlapping, and each record identifier will only be inserted into one cluster. If the thresholds are set to $t_l = t_t = 1.0$, then canopy clustering will generate clusters that are the same as the blocks generated by standard blocking.

Each cluster generated by canopy clustering will become one block, and candidate record pairs will be generated from all pairs of record identifiers within each cluster. Similar to the sorted neighbourhood approach, and q-gram and suffix-array based indexing, a specific candidate record pair will likely be generated from several clusters (blocks). However, each unique candidate record pair will only be compared once in the comparison step.

One drawback with the threshold-based canopy clustering approach is that the size of the generated clusters depends upon the distribution of BKVs, the similarity function used and the setting of the two similarity thresholds. Besides setting the two threshold values, there is no explicit way to limit the size of the generated clusters, and therefore it is not possible to limit the number of candidate record pairs generated.

An alternative way of using the two similarity thresholds is to generate clusters using a nearest-neighbour based approach [64]. Similar to the two thresholds t_l and t_t, two nearest-neighbour parameters are required. The first, n_l, is the number of record identifiers whose records r_x have the highest similarity values with the record selected as cluster centroid, r_c. These n_l records are inserted into a canopy cluster C_i in step 4 of the algorithm. Of these n_l records, the n_t (with $n_t \leq n_l$) will then be removed from the set P of all record identifiers in step 5 of the algorithm. This approach requires that the n_l records in the set P that are most similar to the centroid record r_c need to be identified in each iteration of the algorithm.

The advantage of this nearest-neighbour based approach is that each canopy cluster generated will contain n_l record identifiers. The two parameters n_l and n_t allow explicit control of the number of candidate record pairs that will be generated. A drawback of this approach is however that the fixed size of the generated clusters, n_l, might mean that records that are similar to each other are not inserted into the same cluster. This drawback is similar to the drawback of the sorted neighbourhood approach where a fixed window size can lead to true matches not being compared with each other because they are not inserted into the same window. A recent experimental study has shown that threshold-based canopy clustering can achieve better results than the nearest-neighbour based approach, especially as the databases to be matched or deduplicated get larger [64].

4.9 Mapping Based Indexing

The idea behind mapping based indexing techniques is to convert the BKVs into objects that are mapped into a multi-dimensional space, such that the original similarities between BKVs are preserved [151]. Blocks are then generated similar to canopy clustering by inserting similar objects into the same clusters [151].

The distances between strings are calculated using an approximate string comparison function which needs to be a metric distance function (discussed in Sect. 5.1), such as one of the edit distance based functions (see Sect. 5.3 for details). The multi-dimensional space generated by the mapping process is created one dimension after another using a modification of the *FastMap* [105] algorithm called *StringMap* [151]. StringMap has a linear complexity in the number of strings that are mapped into the multi-dimensional space. The set of strings this algorithm works on is the set of all BKVs. Mapping based indexing consists of two phases.

- In the first phase, the algorithm iterates over the number of dimensions d (which needs to be chosen by the user). For each dimension, StringMap selects two *pivot strings* which ideally are far apart from each other measured using the selected string similarity function. These two strings are used to form the orthogonal directions of the multi-dimensional Euclidean space that is generated.

  An iterative farthest-first algorithm can be used to find the two pivot strings [151]. This algorithm starts by randomly selecting a string, and then finding the string that is farthest away from the first string. In the second iteration, the string in the set of all strings that is farthest away from the second string is found. This process is repeated several times. The last two strings found will be selected as the two pivot strings.

  Once two pivot strings have been selected for a dimension, the coordinates of all other strings for this dimension are calculated based on the two pivot strings. The process is repeated until all dimensions d have been created and a d-dimensional object has been generated for each string.

  A crucial parameter required in this algorithm is the dimensionality d of the space that is generated. Experiments by the developers of StringMap have shown that good results can be achieved with $15 \leq d \leq 25$ [151]. A second crucial issue for mapping based indexing is the type of data structure that is used to store the multi-dimensional objects.

  The developers of StringMap have used an R-Tree [151], which is a popular data structure for efficient storage and retrieval of multi-dimensional objects. However, like most tree-based multi-dimensional index data structures, the higher the dimensionality of the objects to be stored is, the less efficient an R-Tree data structure becomes. The reason for this degradation in efficiency is that with increased dimensionality, more subtrees in an R-tree need to be searched to find objects that are similar to a query object. For dimensions larger than 15 to 20, nearly all objects in a tree-based multi-dimensional index need to be accessed when similarity searches are conducted [3]. This problem is commonly known as the *curse of dimensionality* [135].

- In the second phase of mapping based indexing, similar to canopy clustering based indexing, groups or clusters of objects (that refer to BKVs) are extracted from the multi-dimensional index, and all records that contain any of these BKVs are inserted into the same block. As with canopy clustering, clusters (and thus blocks) can be generated using a similarity threshold or a nearest-neighbour based approach [64]. It is difficult to predict the size of the generated blocks when using a threshold based approach, because block sizes depend upon the distribution of the objects in the multi-dimensional space. Their distribution depends upon the BKVs, the string distance function employed in the mapping and the dimensionality d of the space generated.

A double-embedding based approach to indexing has recently been proposed [1]. The idea of this approach is to first map the BKVs into a space of k dimensions using the StringMap algorithm [151]. This is followed by a mapping of the objects in the k dimensional space into a lower dimensional space (with $k' < k$ dimensions) using the FastMap algorithm [105]. A binary KD-tree data structure is combined with a nearest neighbour based similarity approach to extract similar objects that are then used to generate blocks [1]. An experimental evaluation by the developer of this technique using names and addresses from a publicly available Canadian voter's database has shown improvements in the run-time of between 30 and 60 % reduction compared to the original StringMap based mapping approach, while at the same time achieving the same matching accuracies [1].

4.10 A Comparison of Indexing Techniques

To illustrate the differences in performance of the indexing techniques described in this chapter, this section presents some experimental results of a recent comparative evaluation [64]. This study was based on three small data sets that have previously been used by the data matching community for comparative evaluations. These and other data sets used in data matching research will be discussed in more detail in Sect. 7.5. Table 4.3 provides basic details of these three data sets. They are all available in the *SecondString* toolkit.[1] The true match status of all record pairs is known in all three data sets.

All indexing techniques described in Sects. 4.4–4.9 (and variations of them) have been implemented in the programming language Python as part of the FEBRL[2] open

---

[1] http://secondstring.sourceforge.net

[2] Available from: https://sourceforge.net/projects/febrl/

Table 4.3 Details of the data sets used for the comparative evaluation presented in Fig. 4.10

Data set name	Description	Task	Number of records	Total number of true matches
Census	Synthetic data generated by the US Census Bureau	Linkage	449 / 392	327
Cora	Bibliographic records of machine learning papers	Deduplication	1,295	17,184
Restaurant	Records from Fodor and Zagat restaurant guides	Deduplication	864	112

source record linkage system [61]. FEBRL will be described in detail in Sect. 10.2.4. The results presented here are based on experiments conducted on an otherwise idle compute server with two 2.33 GHz quad-core CPUs and 16 GB of main memory, running Linux 2.6.32 (Ubuntu 10.04) and using Python 2.6.5. Further results, such as the quality of the candidate record pairs generated, are available in a detailed experimental evaluation [64].

The results presented in Fig. 4.10 show the three basic measures time used, number of candidate record pairs generated, and main memory required. Other measures that are commonly used to evaluate indexing techniques will be discussed in Sect. 7.3. As these results illustrate, there are large variations between the different indexing techniques for all three measures. Not just are there differences between indexing techniques, but also for the same technique when applied to different data sets.

Mapping based indexing is the overall slowest technique, followed by q-gram based indexing, canopy clustering and two variations of suffix-array based indexing. Mapping based indexing also produces a large number of candidate record pairs. The more simpler techniques, such as standard blocking, the sorted neighbourhood based approaches and suffix-array based indexing, are faster and require less memory. A major result of this and other comparative studies [20, 64] is that an important factor for successful indexing is the appropriated definition of blocking or sorting keys. How to automatically generate optimal blocking keys based on supervised learning approaches will be discussed in Sect. 4.12 below.

4.11 Other Indexing Techniques

Besides the indexing techniques (and their variations) covered in Sects. 4.4–4.9, different approaches aimed at reducing the time needed to match two databases or deduplicate a single database have been explored.

An iterative blocking technique has recently been proposed that uses the information gained when records within a block are compared to improve the records contained in other blocks [277]. For example, when two records in a block are compared, the resulting merged record can contain information that leads to the merged

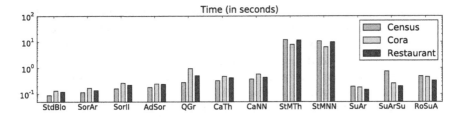

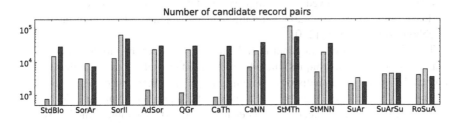

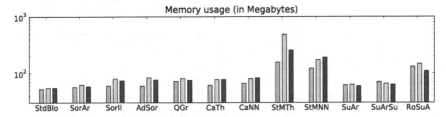

Fig. 4.10 Experimental results for the three small data sets presented in Table 4.3. These results are based on an earlier experimental evaluation by the author [64]. The abbreviations used for the different indexing techniques are: 'StdBlo' refers to the standard blocking technique; 'SorAr' to the array based sorted neighbourhood approach, 'SorII' to the inverted index based sorted neighbourhood approach and 'AdSor' to the adaptive sorted neighbourhood technique; 'QGr' refers to q-gram based indexing; 'CaTh' and 'CaNN' to the threshold and nearest-neighbour based canopy clustering techniques; while 'StMTh' and 'StMNN' refer to the threshold and nearest-neighbour based StringMap indexing techniques; finally, the three suffix-array based approaches are labelled as 'SuAr' (the original suffix array technique), 'SuArSu' (the variation where all substrings are generated), and 'RoSuA' refers to the robust suffix array approach

record being inserted into another block, because the merged record might have a different BKV than the two original records. The issues involved in merging records that have been classified as matches will be covered in Sect. 6.12. The process of indexing a database into blocks (as discussed in Sect. 4.4) with this technique is therefore not static and done only once, but rather blocks are iteratively refined as record pairs are being compared and merged. The experimental evaluation presented by the authors showed that such iterative blocking can be both more accurate and faster at the same time [277].

While all indexing techniques discussed so far only use the attribute values in an individual record to decide into which block to insert the record, a recently developed technique uses the relationship of a record with other records to improve the quality of the indexing process [200]. The idea is to build a relationship graph where records are nodes and relationships between records are vertices. A relationship can for example be between individuals who live at the same address or who share the same telephone number, or between co-authors who have contributed to the same article. Blocks are then generated by randomly selecting a record (similar to canopy clustering and mapping based indexing) and including all records into the block that are connected to this record in the relationship graph. This so-called *semantic blocking* approach achieved much improved matching quality on several data sets while having similar computational requirements as standard blocking or the sorted neighbourhood approach [200].

Techniques that are orthogonal to indexing (i.e. that can be employed complementary to indexing techniques) include running data matching or deduplication on a parallel computer or in a distributed computing environment, where each processing unit will conduct the matching of a subset of the original databases. This topic will be further covered in Sect. 9.5. Related to parallel data matching is the BigMatch technology developed by the US Census Bureau [295]. The challenge faced by this organisation is that very large databases (some containing billions of records) need to be matched with 'smaller' databases (still containing millions of records) on a regular basis. The approach implemented in BigMatch is to load and process the 'smaller' of the two databases into an inverted-index based data structure [288], assuming this data structure fits into the main memory of a large parallel compute server. The indexing step can then be carried out by a single scan through the larger of the two databases while still using several blocking criteria. For each of these blocking criteria, a file will be written that contains plausible matches. Each of these files will be smaller than the original large database. More detailed matching is then carried out separately between the smaller database and each of the generated blocking files [295].

Another approach is to reduce the time required in the expensive detailed comparison of the candidate record pairs that are generated in the indexing step. Comparison functions will be covered in detail in the following chapter. Many of these functions are computationally expensive, so limiting the number of comparisons that need to be conducted can provide substantial performance improvements. Two recently developed techniques [91, 193] explore how an early decision can be made if a candidate record pair is classified as a non-match, and therefore does not need to be compared in detail across many of its attributes. The first proposed approach assumes a distributed environment (such as a crime investigation by a police officer who needs to gather information from several distributed law enforcement databases). A matching tree is developed (similar to a decision tree) that allows an early decision if a candidate record pair will be a match or not. This technique can lead to a significant reduction in communication overhead [91]. A second recent approach assesses the attribute comparisons for a candidate record pair such that a match or non-match decision can be made as early as possible [193]. This is achieved by an optimal ordering of the attributes that are used in the comparison step. Such an optimisation is especially important if the records in the databases to be matched contain many attributes that need to be compared.

Special indexing techniques have recently been proposed for real-time data matching [69, 70], where the aim is to match a stream of query records that contain entity information with a large database that contains records of known entities, such as known criminals or people that have a bad credit history. If the matching of a query record is required in (near) real-time, such as for a police investigation for example, then the amount of permitted matching time for each query record is limited. The matching task then becomes similar to the task of Web search which is successfully carried out by various commercial search engines on very large data collections. The topic of real-time matching will be further discussed in Sect. 9.3.

A large body of work related to indexing has also been conducted by the database community. The two areas relevant to indexing are similarity joins and uncertain or probabilistic databases. The objective of a similarity join is to enable an efficient and scalable approximate join function that calculates the similarities between attribute values in two database tables using a string similarity function such as edit distance or Jaccard distance (to be discussed in the following chapter). Various techniques have been developed that facilitate similarity joins within database environments [22, 127, 170, 232, 272, 289]. The area of uncertain or probabilistic databases is concerned with information that is probabilistic in nature and thus has some uncertainty attached to each value [2]. Indexing techniques for uncertain data have been developed based on standard inverted index and tree-based approaches [216, 298]. Finding similarities between objects in probabilistic databases has recently also received some attention [28]. None of these techniques has however been directly applied in the domain of data matching.

The information retrieval community, besides developing techniques to improve inverted index based indexing for Web search [21], has also investigated techniques to allow efficient detection of duplicate documents returned by Web search engines [131]. Because Web documents are commonly much larger than the records used in data matching and contain more detailed information, a main challenge is to extract relevant parts of documents to allow efficient indexing for scalable duplicate detection [139].

4.12 Learning Optimal Blocking Keys

The blocking or sorting keys that are required for all indexing techniques presented in this chapter are traditionally being defined manually by data matching and domain experts. As experimental evaluations have shown [20, 64], finding optimal blocking key definitions that lead to high quality matching results is challenging. The aim of a good blocking key definition is to get a set of candidate record pairs that includes as many true matches as possible while at the same time keeping the total set of candidate record pairs as small as possible. An alternative way to manually define blocking keys is to learn them automatically from the data that are to be matched.

Two machine learning based approaches to define optimal blocking keys have recently been proposed [34, 188]. Both work in a supervised learning fashion and require training data in the form of record pairs that correspond to true matches and true non-matches. The learning process generates candidates of blocking keys, and using those training examples the candidates that achieve the highest coverage and highest accuracy are selected. The two measures coverage and accuracy are commonly used to evaluate rule-based classification approaches [135]. Coverage measures the number of true matching candidate record pairs in the training set that are covered by a blocking key definition, while accuracy measures how many of the candidate record pairs in the training set that are covered by a blocking key definition correspond to true matches.

The approach developed by Michelson et al. [188] learns so-called *blocking schemes* (blocking criteria) made of {*method*, *attribute*} tuples, where the *method* is a function which compares the values of the given *attribute* from two records. Example methods are *first-1-match*, which returns true if the first characters in the attribute values of two records are the same and false otherwise; *first-3-match*, which returns true if the first three characters are the same, or *token-match*, which returns true if two attribute values have at least one token (word or q-gram) in common. The actual learning algorithm is based on a variation of the sequential covering algorithm [135]. This algorithm learns one rule (one blocking scheme) at a time. It starts with a set of candidate blocking schemes that includes all possible *methods* applied on all available *attributes*. The coverage and accuracy of each possible {*method*, *attribute*} tuple is then calculated using the record pairs in the training data. Individual tuples are combined into conjunctions to improve the accuracy of the learned blocking schemes. The conjunction of {*method*, *attribute*} tuples with the highest coverage and accuracy is selected, and the record pairs that are covered by that conjunction of tuple are removed from the training data set [188]. The process is continued until all training record pairs are covered by a tuple.

An approach similar to learning optimal blocking keys was presented by Bilenko et al. [34]. Instead of learning one rule at a time using the sequential set covering algorithm, blocking schemes are learned by solving an optimisation problem that was shown to be equivalent to the red-blue set cover problem [34]. Both this and the technique proposed by Michelson et al. [188] can in principle be employed with any of the indexing techniques presented in this chapter.

4.13 Practical Considerations and Research Issues

The most important aspects to consider in the indexing step for a practical data matching or deduplication exercise are how to define the blocking key or keys, and what indexing technique to employ. The selection of which attributes to use in a blocking key definition depends upon the number of unique values of an attribute, their frequency distribution and also how many records have an empty value in an attribute. These basic statistics can be gathered through data profiling or data exploration tools, as was previously discussed in Chap. 3. Ideally, attributes that have no missing values and that have a nearly uniform frequency distribution of their values are preferred. The reason for this is that using such attributes will result in blocks or clusters that are of similar sizes, compared to when the values of an attribute follow for example a Zipf-like frequency distribution [64].

The number of blocks and the distribution of their sizes can be further influenced by the use of a (phonetic) encoding function during the generation of BKVs. The choice of a phonetic encoding function depends upon the language of the values that will be used to generate the BKVs. Most phonetic encoding functions, including those presented in Sect. 4.3, have been developed for English names only. For any database that contains either name values from a language other than English, or that contains multilingual name values, the use of phonetic encoding functions should be carefully evaluated, and variations of such functions (that have been appropriately adapted to a certain language) should be considered.

Which indexing technique to use for a certain data matching or deduplication project is a second important practical aspect that needs to be carefully considered. With different indexing techniques, there is usually a trade-off between how many candidate record pairs are generated and the resulting quality of the achieved matching. The more record pairs are removed by an indexing technique from the set of all possible pairs, the more likely some true matching pairs will be removed as well. For databases that contain data of low quality, employing an indexing technique that inserts records into several blocks or clusters will be of advantage compared to employing a technique that inserts each record into one block only. On the other hand, if the data to be matched or deduplicated are of good quality, then using the traditional blocking technique (that inserts each record into one block only) might be appropriate.

One avenue of research in the area of indexing could tackle the challenge of developing multilingual phonetic encoding functions that can be applied on databases that contain names from different languages and cultures. Another area of research is the development of indexing techniques that are scalable, while at the same time also highly efficient. This means that ideally the number of candidate record pairs that are generated only increases linearly with the size of the databases that are matched, while a high matching quality is still achieved by keeping all (or a very high portion of) true matching record pairs in the generated candidate record pairs.

All the indexing techniques that have been presented in this chapter are heuristic approaches. Their aim is to split the records in a database (or databases) into blocks or clusters (that potentially overlap) in such a way that all records that match with each other are inserted into the same block, and records that are not matching are inserted into different blocks. An ultimate goal of research on indexing for data matching is the development of techniques that generate blocks such that it can be proven that (1) all comparisons between records within a block will have a certain minimum similarity with each other (according to some similarity metric), and (2) the similarity between records in different blocks is below this minimum similarity. Specifically, if r_i and r_j are two records, and $sim(r_i, r_j)$ is a similarity measure (such as one of the techniques described in the following chapter) applied to a pair of records (with

$sim(r_i, r_j) = 1$ if $r_i = r_j$ and $sim(r_i, r_j) = 0$ if r_i and r_j are totally different from each other), then an optimal indexing technique would generate blocks B such that $sim(r_i, r_j) \geq t, \forall r_i, r_j \in B_k$, and $sim(r_i, r_j) < t, \forall r_i \in B_k, r_j \in B_l, B_k \neq B_l$, for some threshold $0 \leq t \leq 1$.

4.14 Further Reading

A recent survey by the author arguably provides the most comprehensive comparison of indexing techniques for data matching presented so far [64]. The survey analyses the computation complexity of different indexing techniques, measured as the number of candidate record pairs that are expected to be generated if certain data distributions are assumed. A detailed experimental evaluation on both synthetic and real-world data sets is also provided in this survey. These experiments are highlighting the significant differences in both performance and accuracy that are achieved by various indexing techniques.

Two publications discuss implementation details of how indexing techniques can be employed in industry to match large databases. A TF-IDF based approach, that is similar to canopy clustering described in Sect. 4.8, was used by Koudas et al. in an SQL database environment to match customer information [170]. Various enhancements were presented that can lead to significant performance improvements when large databases are matched. More recently, Weis and Naumann discuss extensions to the sorted neighbourhood approach to allow the scalable deduplication of a customer relationship database containing records in XML format for more than 60 million individuals [272]. The recent introductory book to duplicate detection by the same two authors also contains a discussion of several indexing techniques [195].

Chapter 5
Field and Record Comparison

5.1 Overview and Motivation

As was discussed in Chap. 3, the data used in data matching can be of low quality. They can contain errors and (typographical) variations, name and address values can change over time, and for many personal and other names several valid forms can exist. While this is mostly of concern for attributes such as personal names and addresses, as the examples in Sect. 1.4.6 and 1.4.7 showed, low data quality is also a critical issue for other types of data, including bibliographic databases and consumer product descriptions. Even sophisticated data cleaning and standardisation techniques are not always able to create high quality data that will convert values into exactly the same form for all attributes in pair of records that refer to true matches.

Rather than comparing attribute values between two records using only an exact comparison function (that returns a binary 'same' or 'different' value), it is vital for data matching to employ comparison functions that return some indication of how similar two attribute values are. Such comparison functions need to be tailored to the type of data that are being compared.

Assume $s = sim(a_i, a_j)$ is a similarity function that calculates the numerical similarity s between two attribute values a_i and a_j, where a_i and a_j can be strings, numbers, dates, times, ages, geographic locations, or even more complex values such as text, XML documents, or even multimedia data. It is assumed that such a similarity function sim generates a normalised similarity value s between $0 \leq s \leq 1$. The general requirements of such a function are:

- $sim(a_i, a_i) = 1$: The result of comparing a value with itself is an exact similarity.
- $sim(a_i, a_j) = 0$: The similarity of values that are completely different from each other is 0. What accounts for 'complete different' depends upon the type of data that are compared.
- $0 < sim(a_i, a_j) < 1$: An approximate similarity between exact similarity and total dissimilarity is calculated if two attribute values are somewhat similar to each other. What accounts for 'somewhat similar' again depends upon the type of data that are compared.

P. Christen, *Data Matching*, Data-Centric Systems and Applications,
DOI: 10.1007/978-3-642-31164-2_5, © Springer-Verlag Berlin Heidelberg 2012

There is a correspondence between a similarity function and the mathematical concept of a distance function [135]. A distance function or distance metric $\text{dist}(o_i, o_j)$ between two points (or data objects) o_i and o_j must fulfil the four requirements:

1. $\text{dist}(o_i, o_i) = 0$: The distance from an object to itself is zero.
2. $\text{dist}(o_i, o_j) \geq 0$: The distance between two objects is a non-negative number.
3. $\text{dist}(o_i, o_j) = d(o_j, o_i)$: The distance between two objects is symmetric.
4. $\text{dist}(o_i, o_j) \leq \text{dist}(o_i, o_k) + \text{dist}(o_k, o_j)$: The triangular inequality must hold. It states that the direct distance between two objects is never larger than the combined distance when going through a third object o_k.

To convert a distance value d into a similarity value s, one can either calculate the similarity as $s = \frac{1.0}{d}$, assuming $d > 0$, or as $s = 1 - d$, assuming the distance value is normalised between $0 \leq d \leq 1$.

As will be discussed throughout this chapter, not all similarity comparison functions used for data matching are fulfilling the requirements of a distance function. Some similarity functions are not symmetric, for example those that calculate the inclusion of one attribute value in another. Other measures do not fulfil the triangular inequality. The remainder of this chapter presents a variety of comparison functions that are used in data matching.

5.2 Exact, Truncate and Encoding Comparison

The simple exact comparison of two attribute values (that are assumed to be strings) s_1 and s_2 calculates an exact similarity only:

$$\text{sim}_{\text{exact}}(s_1, s_2) = \begin{cases} 1.0 & \text{if } s_1 = s_2, \\ 0.0 & \text{if } s_1 \neq s_2. \end{cases} \tag{5.1}$$

For string attribute values, there are two variations of the exact comparison function. In the first variation, only the beginning (or end) of the two attribute values are considered for exact comparison. If the first x characters of a string value are denoted with $s[1:x]$ and the last y characters with $s[y:n]$ (with n the number of characters in a string) then two 'truncate' comparison functions can be defined as:

$$\text{sim}_{\text{truncate\_begin(x)}}(s_1, s_2) = \begin{cases} 1.0 & \text{if } s_1[1:x] = s_2[1:x], \\ 0.0 & \text{if } s_1[1:x] \neq s_2[1:x]. \end{cases} \tag{5.2}$$

$$\text{sim}_{\text{truncate\_end(y)}}(s_1, s_2) = \begin{cases} 1.0 & \text{if } s_1[y:n] = s_2[y:n], \\ 0.0 & \text{if } s_1[y:n] \neq s_2[y:n]. \end{cases} \tag{5.3}$$

A second variation of exact string comparison is to first encode the string values s_1 and s_2 using a phonetic (or otherwise) encoding function, as discussed in Sect. 4.3. These encodings replace the original string values with codes in such a way that sim-

		0	1	2	3	4
			g	**a**	**i**	**l**
0		**0**	1	2	3	4
1	**g**	1	**0**	1	2	3
2	**a**	2	1	**0**	1	2
3	**y**	3	2	1	**1**	2
4	**l**	4	3	2	2	**1**
5	**e**	5	4	3	3	**2**

		0	1	2	3	4	5
			p	**e**	**t**	**e**	**r**
0		**0**	1	2	3	4	5
1	**p**	1	**0**	1	2	3	4
2	**e**	2	1	**0**	1	2	3
3	**d**	3	2	1	**1**	2	3
4	**r**	4	3	2	2	**2**	2
5	**o**	5	4	3	3	3	**3**

Fig. 5.1 Levenshtein edit distance example for two pairs of similar name strings. In the first pair, s_1 = 'gayle' and s_2 = 'gail', while in the second pair s_1 = 'pedro' and s_2 = 'peter'. The bold numbers show the paths to the final results. The bottom right corner of each matrix corresponds to the edit distance between each pair of strings

ilar (sounding) strings are replaced with the same code. Assuming that $encode(s)$ is the encoding function, the resulting 'encode' comparison functions can be defined as

$$\text{sim}_{\text{encode}}(s_1, s_2) = \begin{cases} 1.0 & \text{if } encode(s_1) = encode(s_2), \\ 0.0 & \text{if } encode(s_1) \neq encode(s_2). \end{cases} \tag{5.4}$$

All exact comparison functions fulfil the four requirements of a distance function as listed in Sect. 5.1.

5.3 Edit Distance String Comparison

Approximate string comparison functions that are based on the concept of edit distance count the smallest number of edit operations that are required to convert one string into another [196]. Different implementations of this concept account for different types of edit operations. The number of edits between two strings is a distance which can be converted into a similarity, as will be detailed below.

The basic edit distance, also known as *Levenshtein* edit distance [196], is defined as the smallest number of single character insertions, deletions and substitutions that are required to convert one string into another. In its basic form, each edit has the same unit cost 1 associated with it.

Using a dynamic programming algorithm [152], the distance (number of edits) between two strings s_1 and s_2 can be calculated in time $O(|s_1| \times |s_2|)$ using $O(\min(|s_1|, |s_2|))$ space, with $|\cdot|$ denoting the length of a string in characters. Figure 5.1 shows two string pairs and the corresponding matrices d that are used to calculate the edit distance between them. A cell $d[i, j]$ in row i ($0 \leq i \leq |s_1|$) and column j ($0 \leq j \leq |s_2|$) in these matrices corresponds to the number of edits required to convert the first i characters of string s_1 (shown in the first column of a matrix) into the string comprised of the first j characters of string s_2 (shown in the top row of a matrix). For example, cell $d[4, 2]$ in the left matrix (with values 2) in Fig. 5.1 corresponds to the number of edits required to convert string 'gayl' into 'ga' (two character deletions).

The dynamic programming algorithm starts by filling in the first row and first column of the matrix with the corresponding column or row values. The cell $d[0, j]$ in row 0 and column j $(0 \leq j \leq |s_1|)$ is filled with the value j, and the cell $d[i, 0]$ in row i and column 0 is filled with the value i $(0 \leq i \leq |s_2|)$. The remaining cells of the matrix are filled using the following recursive approach:

- If $s_1[i] = s_2[j]$, then

$$d[i, j] = d[i - 1, j - 1].$$

- If $s_1[i] \neq s_2[j]$, then

$$d[i, j] = \text{minimum} \begin{cases} d[i - 1, j] + 1 & \text{a deletion,} \\ d[i, j - 1] + 1 & \text{an insertion, or} \\ d[i - 1, j - 1] + 1 & \text{a substitution.} \end{cases} \tag{5.5}$$

As Eq. 5.5 shows, the dynamic programming approach used to calculate the Levenshtein edit distance only requires two rows of the matrix d to be stored at any time [196], and therefore $O(\min(|s_1|, |s_2|))$ space is needed. The final Levenshtein edit distance between s_1 and s_2 is the value in the lower right corner cell, $\text{dist}_{\text{levenshtein}}(s_1, s_2) = d[|s_1|, |s_2|]$. It can be converted into a similarity (between 0.0 and 1.0) using

$$\text{sim}_{\text{levenshtein}}(s_1, s_2) = 1.0 - \frac{\text{dist}_{\text{levenshtein}}(s_1, s_2)}{\max(|s_1|, |s_2|)}. \tag{5.6}$$

The Levenshtein edit distance is symmetric with respect to s_1 and s_2, and it always holds that $0 \leq \text{dist}_{\text{levenshtein}}(s_1, s_2) \leq \max(|s_1|, |s_2|)$. The absolute difference in the lengths of two strings is also a lower bound for the Levenshtein edit distance between them: $\text{abs}(|s_1| - |s_2|) \leq \text{dist}_{\text{levenshtein}}(s_1, s_2)$. This property allows quick filtering of string pairs that have a large difference in their lengths without the need to fully calculate the edit distance between them [128].

A variation of the Levenshtein edit distance, called the Damerau-Levenshtein edit distance, adds as a fourth basic edit operation the transposition (swapping) of two adjacent characters [89, 196]. Transpositions are common typing errors, such as, for example, the variation 'Sydeny' for the city name 'Sydney' (a favourite typing error of the author). The Damerau-Levenshtein edit distance, $\text{dist}_{\text{damerau\_levenshtein}}(s_1, s_2)$, of two strings s_1 and s_2 is always smaller or equal to the original Levenshtein edit distance of the same pair of strings

$$\text{dist}_{\text{damerau\_levenshtein}}(s_1, s_2) \leq \text{dist}_{\text{levenshtein}}(s_1, s_2).$$

As a result, the Damerau-Levenshtein similarity, calculated in a similar way as the Levenshtein similarity in Eq. (5.6) [89], is always equal to or larger than the Levenshtein similarity

$$\text{sim}_{\text{damerau\_levenshtein}}(s_1, s_2) \geq \text{sim}_{\text{levenshtein}}(s_1, s_2).$$

Various improved algorithms have been proposed to reduce the quadratic complexity of the basic Levenshtein edit distance algorithm and make it more efficient for comparing longer strings [196]. Other extensions of the basic edit distance algorithm allow different costs for the different edit operations [133], for example, a cost of 1.0 for insertions and deletions, and 0.5 for substitutions. These costs can be used in the recursive calculations in Eq. (5.5). This can be useful where any change in the lengths of a string is seen as a more severe error compared to a character change. Postcodes or zipcodes can be types of data of where this could be appropriate. The interested reader can refer to surveys of edit distance based approximate string comparison functions [152, 196] for more detailed information.

It is even possible to have different costs for edits on different individual characters. For example, a substitution from letter 'q' to 'g' might be given a smaller edit cost compared to the cost of substituting 'x' to 'i', because the visual similarity of the first character pair is much higher than for the second pair. Such an extension requires a transformation table, where each character pair has an associate cost listed. These costs can be based on character shape (as shown above) to take care of optical character recognition (OCR) errors, for example, or depend upon how similar sounding two characters are (as discussed in Sect. 4.3 on phonetic encodings).

Extending on the idea of assigning specific costs to edits of individual character pairs, several techniques have been developed in recent years which learn the optimal costs of edits from training data [35, 84, 293, 300]. For such approaches, pairs of strings are required that correspond to known true matches and non-matches, i.e. pairs that refer to the same underlying entity or to two different entities (this will be covered in more detail in Sect. 6.6). The frequencies of certain edits occurring in matching and non-matching string pairs are used to calculate the costs associated with an edit. An edit that occurs more frequently in non-matching pairs will have a higher cost than an edit that is more frequent in matching pairs (and is thus deemed to be a valid variation between two string values).

5.3.1 Smith-Waterman Edit Distance String Comparison

Another edit distance based approximate string comparison technique is the Smith-Waterman edit distance [84, 191]. This algorithm was originally developed to find the optimal alignment between biological sequences, such as DNA or protein sequences. It is also based on a dynamic programming approach similar to the Levenshtein edit distance, but it allows for gaps as well as character specific match scores or costs.

This algorithm has previously been employed in the domain of data matching, and in the following description the parameter values (scores) presented by Monge and Elkan [191] are used. The Smith-Waterman edit distance has five basic operations, each with a different match score, ms:

- $ms_m = 5$: An exact match between two characters.

- $ms_s = 2$: An approximate match between two similar characters. Character similarity can be based on letter groupings such as the ones generated from the Soundex encoding (Fig. 4.2). Similar character pairs would then, for example, be 'd' and 't' or 'm' and 'n'.
- $ms_d = -5$: A mismatch between two different characters (that are neither equal nor similar).
- $ms_g = -5$: A gap start penalty, where there is at least one character in one string that does not appear in the other string.
- $ms_c = -1$: A gap continuation penalty, where a previously started gap (missing character) continues.

The final, overall best score, $bs_{\text{smith\_waterman}}$ is the highest value (rather than the lowest as with the Levenshtein and Damerau-Levenshtein edit distances) within the dynamic programming score matrix. From this best score a similarity value $\text{sim}_{\text{smith\_waterman}}(s_1, s_2)$ can be calculated using:

$$\text{sim}_{\text{smith\_waterman}}(s_1, s_2) = \frac{bs_{\text{smith\_waterman}}}{\text{div}_{\text{smith\_waterman}} \times ms_m}, \qquad (5.7)$$

with ms_m the value when two characters match, and $\text{div}_{\text{smith\_waterman}}$ a factor that can be calculated in one of three ways:

- $\text{div}_{\text{smith\_waterman}} = \min(|s_1|, |s_2|)$,
- $\text{div}_{\text{smith\_waterman}} = \max(|s_1|, |s_2|)$, or
- $\text{div}_{\text{smith\_waterman}} = \frac{|s_1| + |s_2|}{2}$.

The first factor corresponds to the overlap coefficient and the third factor to the Dice coefficient, respectively, as will be discussed in the following section.

As it allows for gaps, the Smith-Waterman edit distance can, for example, be suitable for compound names that contain initials or abbreviated names. A major drawback of the algorithm is, however, its computation complexity. For calculating the distance between two strings s_1 and s_2, the algorithm requires $O(|s_1| \times |s_2|)$ space, and its time complexity is $O(\min(|s_1|, |s_2|) \times |s_1| \times |s_2|)$. Various improvements to the basic Smith-Waterman algorithm have been developed which reduce the time complexity to $O(|s_1| \times |s_2|)$ [196]. Many of these improved algorithms have been developed by the bioinformatics community, where very long sequences (of genomes and proteins) need to be compared.

5.4 Q-gram Based String Comparison

The idea behind q-gram based approximate string comparison is to split the two input strings into short sub-strings of length q characters (called q-grams) using a sliding window approach, and to count how many of these q-grams occur in both input strings. Starting from the first position, q characters are selected into the first q-gram, then starting from the second position q characters are selected into the

String	Bigrams	Padded bigrams	Positional bigrams	Trigrams
gail	ga, ai, il	⊙g, ga, ai, il, l⊗	(ga,1), (ai,2), (il,3)	gai, ail
gayle	ga, ay, yl, le	⊙g, ga, ay, yl, le, e⊗	(ga,1), (ay,2), (yl,3), (le,4)	gay, ayl, yle
peter	pe, et, te, er	⊙p, pe, et, te, er, r⊗	(pe,1), (et,2), (te,3), (er,4)	pet, ete, ter
pedro	pe, ed, dr, ro	⊙p, pe, ed, dr, ro, o⊗	(pe,1), (ed,2), (dr,3), (ro,4)	ped, edr, dro

Fig. 5.2 Example of bigrams ($q = 2$), padded bigrams, positional bigrams and trigrams ($q = 3$) for two pairs of name strings

second q-gram, and so on. The number c of q-grams in a string s equals to $c = |s| - q + 1$, where $|s|$ is the number of characters in the string. Figure 5.2 illustrates this process on two pairs of given name strings. Q-grams are also called n-grams [172]. The most commonly selected values of q for attributes used in data matching, such as names and addresses, is $q = 2$ (called *bigrams* or *digrams* [162]) or $q = 3$ (called *trigrams* [258]).

Once the q-grams of two strings, s_1 and s_2, are generated, the similarity between s_1 and s_2 is calculated based on the number of q-grams the two strings have in common. If c_{comon} denotes the number of q-grams in common between s_1 and s_2, c_1 the number of q-grams in string s_1, and c_2 the number of q-grams in string s_2, then a normalised numerical similarity in the range of $0.0 \leq s \leq 1.0$ can be calculated using one of the following three methods:

$$\text{Overlap coefficient: } \text{sim}_{\text{overlap}}(s_1, s_2) = \frac{c_{\text{common}}}{\min(c_1, c_2)}, \tag{5.8}$$

$$\text{Jaccard coefficient: } \text{sim}_{\text{jaccard}}(s_1, s_2) = \frac{c_{\text{common}}}{c_1 + c_2 - c_{\text{common}}}, \tag{5.9}$$

$$\text{Dice coefficient: } \text{sim}_{\text{dice}}(s_1, s_2) = \frac{2 \times c_{\text{common}}}{c_1 + c_2}. \tag{5.10}$$

An extension of the Jaccard coefficient for multiword strings will be presented in Sect. 5.7.

Taking the first string pair from Fig. 5.2 as an example, the only common bigram is 'ga', and thus $c_{\text{common}} = 1$. The different q-gram based similarity values for this pair are then calculated as:

$$\text{sim}_{\text{overlap}}(\text{'gail'}, \text{'gayle'}) = \frac{1}{3} = 0.333,$$

$$\text{sim}_{\text{jaccard}}(\text{'gail'}, \text{'gayle'}) = \frac{1}{3 + 3 - 1} = \frac{1}{5} = 0.2,$$

$$\text{sim}_{\text{dice}}(\text{'gail'}, \text{'gayle'}) = \frac{2 \times 1}{3 + 4} = \frac{2}{7} = 0.286.$$

For trigrams, the similarity for this pair is 0.0 for all three methods, because no trigram appears in common. The same occurs for the second string pair from Fig. 5.2, illustrating the sensitivity of trigrams to single character differences when comparing short strings.

There are two extensions to the basic q-gram based approach, as illustrated in Fig. 5.2. In the first extension, q-grams are *padded* with $q - 1$ special characters at the beginning and the end of each string before its q-grams are generated. The aim of padding is to provide specific information about the start and end of string values. The special characters should be different from the characters that are expected in an attribute, so the start and end q-grams are different from any q-gram that is generated from the actual string value. In Fig. 5.2, the $\odot$ symbol represents the special start character and $\otimes$ the special end character.

The similarity values calculated with padded q-grams will be larger for strings that have the same beginning and end but errors in the middle, but will lead to lower similarity values if there are different characters at the beginning or end. Empirical studies have shown that padded q-grams can increase data matching quality [162]. The calculation of similarities using padded q-grams is the same as for non-padded q-grams following one of Eqs. (5.8), (5.9) or (5.10).

The second extension, also shown in Fig. 5.2, is to add positional information to q-grams. Each q-gram is given the position number where it occurs within a string. When the number of q-grams in common between two strings is calculated, only q-grams that are the same and that have a position value within a certain maximum distance are considered. This maximum distance can either be an absolute value that is independent of the lengths of the two strings that are compared, or it can be adjusted according to the lengths of the two strings. For example, if the maximum distance is set to 40 % of the average string length of a pair, then for the second pair in Fig. 5.2 this distance would be 2. Therefore, the positional bigram ('pe',1) would only be considered to be common with the other positional bigrams ('pe',1), ('pe', 2) and ('pe', 3), but not with ('pe',4). Once the q-grams in common, c_{common}, have been calculated, then the three similarity measures in Eqs. (5.8), (5.9) or (5.10) can be used to calculate the actual similarity value between two strings. Positional q-grams can also be padded in a way similar to non-positional q-grams.

The computation complexity of q-gram based comparison functions in both time and amount of memory needed is $O(|s_1| + |s_2|)$. This is a much smaller complexity compared to the edit distance based comparison functions, which makes q-gram based string comparisons more efficient especially for longer strings.

A novel algorithm based on *skip-grams* was recently proposed, aimed at improving the matching within a cross-lingual information retrieval system [162]. The basic idea of this approach is to not only form bigrams from two adjacent characters, but also bigrams that skip characters (called *skip-grams*). So-called *gram-classes* are defined to specify the type of skip-grams to be created. For example, for a gram-class $gc = \{0, 1\}$ and the string 'pedro', the following skip-grams are created: 'pe', 'ed', 'dr', 'ro' (0-skip grams, i.e. the normal bigrams) and 'pd', 'er', 'do' (1-skip grams which skip one character). The properties of various gram-classes and how they relate to character edits such as insertions, deletions and substitutions are discussed by the developers of this approach [162]. Their experimental evaluation using multilingual texts from different European languages showed improved matching results com-

String 1	String 2	c	t	p	c_{sim}	sim_{jaro}	$sim_{winkler}$	$sim_{winkler\_long}$	$sim_{winkler\_sim}$
shackleford	shackelford	11	1	4	0	0.9697	0.9818	0.9886	0.9697
nichleson	nichulson	8	0	4	0.3	0.9259	0.9556	0.9667	0.9481
jones	johnson	4	0	2	0.3	0.7905	0.8324	0.8491	0.8248
massey	massie	5	0	4	0.3	0.8889	0.9333	–	0.9222
jeraldine	geraldine	8	0	0	0.3	0.9259	0.9259	0.9519	0.9481
michelle	michael	6	0	4	0.3	0.8690	0.9214	0.9302	0.8958

Fig. 5.3 Example of the Jaro and Winkler string comparison functions for several pairs of name strings. The characters in common in a string pair are shown in bold. The transposed characters in the first name pair are 'le' and 'el'. For the pair 'massey' and 'massie' the Winkler long adjustment is not applied, while for the pair 'jeraldine' and 'geraldine' the Winkler prefix adjustment does not change the basic Jaro similarity value because the first character of these two names is different

pared to bigrams, trigrams, edit distance and the longest common substring (LCS) (discussed in Sect. 5.9) based approximate string comparison techniques.

5.5 Jaro and Winkler String Comparison

This family of approximate string comparison functions was developed by Matthew Jaro and William Winkler from the US Census Bureau [149, 279]. These functions are designed specifically for the comparison of names, and they take various heuristics into account that are based on the experience of data matching conducted over many years at the US Census Bureau.[1]

The basic Jaro comparison function combines edit distance and *q*-gram based comparison techniques. It counts the number of characters that are common in two strings within a certain window of characters, similar to the positional *q*-gram based comparison approach. Specifically, the Jaro function counts the number of agreeing characters c (characters that are the same) that are in common within half the length of the longer string, and the number of transpositions t (two adjacent characters that are swapped in the two strings, such as 'pe' and 'ep') in the sets of common strings. Based on these two counts, the Jaro similarity value is calculated as [294]:

$$\text{sim}_{\text{jaro}}(s_1, s_2) = \frac{1}{3}\left(\frac{c}{|s_1|} + \frac{c}{|s_2|} + \frac{c-t}{c}\right). \tag{5.11}$$

Several modifications have been introduced to the Jaro algorithm based on experiences from data matching projects conducted by the US Census Bureau [294], as well as empirical studies that have found that fewer errors appear at the beginning of names compared to the middle or end [214].

---

[1] It is interesting to note that until around ten years ago no references to the Jaro or Winkler string comparators could be found in the computer science literature, although they have been developed and used in the matching of census data since the 1980s

The first modification is to increase the similarity between two strings if their beginning is the same and differences only occur toward the middle and end of the two strings. The basic Winkler algorithm increases the Jaro similarity value for up to four agreeing initial characters. The basic Winkler similarity is calculated as [294]:

$$\text{sim}_{\text{winkler}}(s_1, s_2) = \text{sim}_{\text{jaro}}(s_1, s_2) + (1.0 - \text{sim}_{\text{jaro}}(s_1, s_2))\frac{p}{10}, \qquad (5.12)$$

with p ($0 \leq p \leq 4$) being the number of agreeing characters at the beginning of two strings (common prefix). For example, for the name pair 'peter' and 'petra' the common prefix is 'pet', and therefore $p = 3$.

The second modification further adjusts the similarity value for strings that are both at least 5 characters long, and that have at least two common characters besides the common prefix p [294]. If c is the number of common characters and p the number of common characters in the prefix, then the conditions required to employ this modification are:

$$\min(|s_1|, |s_2|) \geq 5,$$
$$c - p \geq 2,$$
$$c - p \geq \frac{\min(|s_1|, |s_2|) - p}{2}.$$

The third condition requires that besides the common prefix the two strings have at least half of the remaining characters of the shorter string in common. If all three conditions hold, then the similarity between two strings will be adjusted

$$\text{sim}_{\text{winkler\_long}}(s_1, s_2) = \text{sim}_{\text{winkler}}(s_1, s_2) \qquad (5.13)$$
$$+ (1.0 - \text{sim}_{\text{winkler}}(s_1, s_2))\frac{c - (p + 1)}{|s_1| + |s_2| - 2(p - 1)}.$$

A third modification, that is orthogonal to the previous two, is to adjust the similarity value if the pair of strings contains characters that are similar, and thus are more likely to be substituted with each other in misspellings [294]. A set of 36 character pairs has, for example, been implemented in the BigMatch program of the US Census Bureau [295]. Examples of such pairs include ('a','e'), ('e','u'), ('w','v'), ('s','z') or ('5','s'). This modification works by counting the number of similar characters in the remainder of the two strings after the common characters have been removed. For example, for the name pair 'nichelson' and 'nichulson' from Fig. 5.3, only the character pair 'e' and 'u' is not in common. These two letters are a similar pair, and therefore the similarity is increased.

Similar to the definition of common characters (only occurring within half the string length) of the Jaro comparison function, similar characters are also only searched within a certain distance. For each similar character pair that is found, the count of common characters, c, is increased by 0.3. So if c_s similar pairs have been found, then $c_{\text{sim}} = 0.3 \times c_s$, and the basic Jaro similarity calculation is adjusted to:

$$\text{sim}_{\text{winkler\_sim}}(s_1, s_2) = \frac{1}{3} \left(\frac{(c + c_{\text{sim}})}{|s_1|} + \frac{(c + c_{\text{sim}})}{|s_2|} + \frac{c - t}{c} \right). \qquad (5.14)$$

This similarity adjustment can be combined with the prefix and long string adjustments by modifying Eqs. (5.12) and (5.13) accordingly.

The time and space complexities of both the Jaro and the Winkler algorithms are $O(|s_1| + |s_2|)$.

5.6 Monge-Elkan String Comparison

This approximate string comparison function was developed specifically for calculating the similarity of string values that contain several words [191, 192]. Such values occur commonly in data matching of business names, addresses or where personal names have not been standardised and segmented as discussed in Chap. 3.

The idea of this approach is to first extract the tokens (words or elements separated by whitespace characters) in the two input strings, and then to find the best matching pairs of tokens in the sets of tokens using a secondary similarity function sim'. Specifically, following the notation used in [191], in this recursive matching scheme the two strings, s_1 and s_2, are split into the two sets of tokens, A and B, and the similarity between s_1 and s_2 is calculated as:

$$\text{sim}_{\text{monge\_elkan}}(s_1, s_2) = \frac{1}{|A|} \sum_{i=1}^{|A|} \max_{j=1}^{|B|} sim'(A_i, B_j), \qquad (5.15)$$

with $|A|$ the number of tokens in s_1, $|B|$ the number of tokens in s_2, and sim' a similarity function that calculates the actual numerical similarity value (between 0.0 and 1.0) of two tokens.

The computation complexity of this comparison function is quadratic in the number of tokens, because each token in A needs to be compared with every token in B. If both A and B only contain one token, then the Monge-Elkan comparison function reduces to the secondary similarity function sim'.

As an example, for the two strings 'peter christen' and 'christian pedro' and using the Jaro comparison function as the secondary similarity function, sim', the four calculated similarities are:

$$\text{jaro('peter', 'christian')} = 0.3741$$
$$\text{jaro('peter', 'pedro')} = 0.7333$$
$$\text{jaro('christen', 'christian')} = 0.8843$$
$$\text{jaro('christen', 'pedro')} = 0.4417.$$

The two best matching pairs of tokens (names) are 'peter' with 'pedro', and 'chris-ten' with 'christian', resulting in the final similarity $\text{sim}_{\text{monge\_elkan}}$('peter christen', 'christian pedro')$= \frac{1}{2}(0.7333 + 0.8843) = 0.8088$.

5.7 Extended Jaccard Comparison

The basic Jaccard coefficient [195] (Eq. 5.9) calculates the similarity between two sets of items (or tokens), A and B, as the number of items contained in the intersection of the two sets divided by the number of items contained in the union of the two sets:

$$\text{sim}_{\text{jaccard}}(A, B) = \frac{|A \cap B|}{|A \cup B|} = \frac{|A \cap B|}{|A| + |B| - |A \cap B|}. \quad (5.16)$$

This approach is also used in the q-gram based string comparison function described in Sect. 5.4, where the tokens in A and B are the q-grams extracted from the strings that are compared.

If the strings s_1 and s_2 that are compared contain several words, and thus the tokens to be compared become words rather than q-grams, then the basic Jaccard similarity can be extended in a similar way to the Monge-Elkan approach. A secondary similarity function sim' is used to calculate the similarity between all pairs of tokens (words) in the two strings to be compared. Assuming that A is the set of tokens extracted from string s_1 and B the set of tokens extracted from string s_2, then the set of tokens shared between the two strings, S, is defined as [195]:

$$S = \{(a_i, b_j)|a_i \in A \wedge b_j \in B : sim'(a_i, b_j) \geq \theta\}, \quad (5.17)$$

where θ is the similarity threshold $(0 < \theta < 1)$ that is used to decide if two tokens are similar or not. Pairs of tokens with a similarity value above θ are classified as matching and are included in the set S of shared tokens.

The set of tokens in A that are unique to string s_1 (i.e. that are not in the set S) is denoted with $U_A = \{a_i|a_i \in A \wedge b_j \in B \wedge (a_i, b_j) \notin S\}$, and the set of tokens unique to string s_2 is denoted with $U_B = \{b_j|a_i \in A \wedge b_j \in B \wedge (a_i, b_j) \notin S\}$. An extended Jaccard similarity function can now be defined as [195]:

$$\text{sim}_{\text{jaccard\_ext}} = \frac{|S|}{|S| + |U_A| + |U_B|}. \quad (5.18)$$

A further extension is that weights are assigned to both the shared tokens as well as the unique tokens [195]. The weights $w(a_i, b_j)$ of the shared tokens in S can be set to the similarity values calculated using the secondary similarity function, $w(a_i, b_j) = sim'(a_i, b_j)$, while the weights for the unique tokens can be set to $w(a_i) = w(b_j) = 1.0$. Alternatively, weights can be assigned to the unique tokens according to their importance or relevance. This idea is implemented in SoftTFIDF, the comparison function described in the following section.

5.8 SoftTFIDF String Comparison

An approximate string comparison function that is related to the Monge-Elkan approach has been proposed by Cohen et al. [84]. It is based on ideas that are used in the field of information retrieval and is implemented in the SecondString approximate string comparison library,[2] which will be described in more detail in Sect. 10.2.9.

The concept of using Term Frequency (TF) and inverse document frequency (IDF) has been developed in the field of information retrieval to give weights to terms and documents when calculating the similarities between a query and the documents in a collection, for example, when conducting Web search [288]. Term frequency gives a higher weight to terms that occur more frequently in a document, with the underlying idea that the more frequent a term is in a document the more relevant it is to that document. IDF, on the other hand, gives higher weights to terms that occur less frequent in a document collection, with the idea that terms that are less frequent in a collection are more important to distinguish documents that are more relevant to this term from those documents that are less relevant. TF-IDF combines these two weights into a single numerical value. There are different variations of how TF and IDF can be calculated. The interested reader can refer to the book by Witten et al. [288] for more details.

Each document, D_i, in a collection is represented as a document vector, d_i. The dimensionality of these vectors equals to the number of unique terms in the collection, and each element (dimension) in these vectors corresponds to a unique term in the collection, with the numerical value of the element being the TF-IDF weight of that term in a document.

The similarity between two documents in a collection, or between a query (also represented as document vector) and a document, can be calculated as the Cosine similarity between their two document vectors. Cosine similarity is the angle between two vectors in the high-dimensional space generated by the two vectors. Assuming there are n unique terms in a document collection, and $d_i = [w_{i,1}, w_{i,2}, \ldots, w_{i,n}]$ is the vector of document D_i and $d_j = [w_{j,1}, w_{j,2}, \ldots, w_{j,n}]$ the vector of document D_j, with $w_{d,t}$ the TF-IDF weight of term t in document d, then the Cosine similarity is calculated as [288]:

$$\text{sim}_{\text{cosine}}(D_i, D_j) = \frac{1}{W_i W_j} \sum_{t=1}^{n} w_{i,t} \cdot w_{j,t}, \tag{5.19}$$

where

$$W_i = \sqrt{\sum_{t=1}^{n} w_{i,t}^2} \quad \text{and} \quad W_j = \sqrt{\sum_{t=1}^{n} w_{j,t}^2}.$$

---

[2] See: http://secondstring.sourceforge.net/

Applying this idea to records used in data matching, which are usually much shorter than the documents used in Web search or in digital libraries, the SoftTFIDF comparison function is defined similar to the Monge-Elkan comparison function. Assuming A and B are the two sets of tokens in the strings s_1 and s_2, and a secondary similarity function sim' is used to calculate the similarities between individual pairs of tokens (or words). Let CLOSE(θ, A, B) denote the set of tokens $a_i \in A$ such that there is some token $b_j \in B$ that fulfils $sim'(a_i, b_j) \geq \theta$, and for $a_i \in$ CLOSE(θ, A, B) let $N(a_i, B) = \max(\{sim'(a_i, b_j)|b_j \in B\})$ [192]. The SoftTFIDF similarity is then defined as [84]:

$$\text{sim}_{\text{softtfidf}}(s_1, s_2) = \sum_{u \in \text{CLOSE}(\theta, A, B)} V(u, A) \cdot V(u, B) \cdot N(u, B), \qquad (5.20)$$

where $V(u, A)$ is the TF-IDF weight of token u in token-set A and $V(u, B)$ the TF-IDF weight of token u in token-set B. In the experiments presented by the authors of SoftTFIDF [84], the Winkler string comparison function was used to calculate sim', with a threshold $\theta = 0.9$.

One difficulty with SoftTFIDF is that the weights $V(u, A)$ and $V(u, B)$ need to be calculated over the databases that are to be matched, requiring full scans of these databases and storing these weights for all unique tokens that occur in the attributes used from both databases. A recent comparison of the Monge-Elkan and SoftTFIDF similarity functions highlighted that SoftTFIDF can potentially result in a similarity value larger than 1.0, because the approximate matching does not prevent a token $b_j \in B$ to be matched with more than one token $a_i \in A$.

5.9 Longest Common Substring Comparison

The idea of this algorithm is to find and remove the longest substring s_c that two strings s_1 and s_2 have in common in an iterative fashion, as long as the common substring found contains a minimum number of characters, $l_{\min}$ [113]. A substring is defined as a consecutive sequence of characters.

The Longest Common Sub-String (LCS) algorithm starts by identifying the longest substring, s_{c_1}, that s_1 and s_2 have in common. This common substring is then removed from the two input strings, resulting in two shorter strings s_1' and s_2'. The process is repeated, by finding the longest substring s_{c_2} that s_1' and s_2' have in common, and removing this common substring. The process stops when a substring s_{c_n} is found that contains less than $l_{\min}$ characters. The value for $l_{\min}$ is commonly set to $l_{\min} = 2$ or $l_{\min} = 3$.

The total summed length of all found common substrings, $l_c = \sum_{i=1}^{n} |s_{c_i}|$, with n the number of common substrings found, is then used to calculate a similarity value between 0.0 and 1.0 using either the Overlap, Jaccard or Dice coefficient (similar to Eqs. (5.8)–(5.10)):

$$\text{sim}_{\text{lcs\_overlap}}(s_1, s_2) = \frac{l_c}{\min(|s_1|, |s_2|)}, \tag{5.21}$$

$$\text{sim}_{\text{lcs\_jaccard}}(s_1, s_2) = \frac{l_c}{|s_1| + |s_2| - |l_c|}, \tag{5.22}$$

$$\text{sim}_{\text{lcs\_dice}}(s_1, s_2) = \frac{2 \times l_c}{|s_1| + |s_2|}. \tag{5.23}$$

With the example previously used in Sect. 5.6, $s_1 =$ 'peter christen' and $s_2 =$ 'christian pedro', and with a minimum common length $l_{\min} = 2$, the LCS algorithm in the first step will find the LCS $s_{c_1} =$ 'christ', leaving the two strings $s_1' =$ 'peter en' and $s_2' =$ 'ian pedro'. In the second iteration, the LCS $s_{c_2} =$ 'pe', leaving the two strings $s_1'' =$ 'ter en' and $s_2'' =$ 'ian dro'. No more common substring is found, therefore $l_c = |\text{'christ'}| + |\text{'pe'}| = 6 + 2 = 8$, resulting in LCS similarities $\text{sim}_{\text{lcs\_overlap}} = \frac{8}{14} = 0.5714$, $\text{sim}_{\text{lcs\_jaccard}} = \frac{8}{14+15-8} = 0.381$, and $\text{sim}_{\text{lcs\_dice}} = \frac{2 \times 8}{14+15} = 0.5517$.

The LCS string comparison function is well suited for strings that contain several words that potentially are not in the same order (such as given name and surname values that are swapped). For example, the string pair 'peter christen' and 'christen peter' have $s_{c_1} =$ 'christen' and $s_{c_2} =$ 'peter' in common (and so $l_c = 13$ out of the 14 characters in these two strings), and because $|s_1| = |s_2|$ both Eqs. (5.21) and (5.23) will return $\text{sim}_{\text{lcs\_overlap}} = \text{sim}_{\text{lcs\_dice}} = \frac{13}{14} = 0.929$, while Eq. (5.22) will return $\text{sim}_{\text{lcs\_jaccard}} = \frac{13}{(14+14-13)} = \frac{13}{15} = 0.867$.

The implementation of the LCS string comparison function follows a dynamic programming algorithm [113] with a time complexity of $O(|s_1| \times |s_2|)$ and using $O(\min(|s_1|, |s_2|))$ space. Because this dynamic programming approach starts from the beginning of the strings, the resulting LCS similarity might not be symmetric if the two input strings are swapped. For example, for the string pair $s_1 =$ 'prap' and $s_2 =$ 'papr' and $l_{\min} = 2$, $s_{c_1} =$ 'pr' will be selected, resulting in $s_1' =$ 'ap' and $s_2' =$ 'pa'. As no more common substrings of minimum length 2 are available in s_1' and s_2', the similarity of s_1 and s_2 is $\text{sim}_{\text{lcs}} = 0.5$. On the other hand, if the strings are swapped, $s_1 =$ 'papr' and $s_2 =$ 'prap', then $s_{c_1} =$ 'ap' will be selected, resulting in $s_1' =$ 'pr' and $s_2' =$ 'pr', and therefore a second common substring of length 2, $s_{c_2} =$ 'pr'. The total length of the two-substrings is $l_c = 2 + 2 = 4$, and thus $\text{sim}_{\text{lcs}} = 1.0$. To overcome this situation, the LCS algorithm needs to be run twice with the input strings swapped and the calculating similarities averaged.

A generalisation of the longest substring algorithm is to consider longest common sub-sequences as well [27]. While a substring is a sequence of consecutive characters, a sub-sequence consists of characters that are not necessarily consecutive. For example, the string 'peter' is a sub-sequence of the string 'pedro foster' (the underlined characters), while it is not a substring. A substring of a longer string is always a sub-sequence of that string, however, a sub-sequence of a string is not always a substring.

5.10 Other Approximate String Comparison Techniques

Various other techniques for approximate string comparison have been developed. Some of them are specific to certain domains (such as personal names or biomedical sequences), while others assume values are from a certain language. This section briefly covers a few techniques that have been used in the domain of data matching.

5.10.1 Bag Distance

The *bag distance* has been proposed as a computationally cheap approximation to edit distance [18]. A bag x is defined as the multi-set ms of the characters in a string s, $x = ms(s)$. A multi-set contains a character as many times as it occurs in the original string. For example, if s_1 = 'peter' then $x = ms(s_1) = \{e, e, p, r, t\}$, while for s_2 = 'pedro' then $y = ms(s_2) = \{d, e, o, p, r\}$. The bag distance is then defined as:

$$\text{dist}_{\text{bag}} = \max(|x - y|, |y - x|). \tag{5.24}$$

Continuing the above example, $|x - y| = |\{e, t\}| = 2$ and $|y - x| = |\{d, o\}| = 2$, and therefore $\text{dist}_{\text{bag}}(s_1, s_2) = 2$. Converting bag distance into a normalised numerical similarity is then done based on Eq. (5.6):

$$\text{sim}_{\text{bag}}(s_1, s_2) = 1.0 - \frac{\text{dist}_{\text{bag}}(s_1, s_2)}{\max(|s_1|, |s_2|)}. \tag{5.25}$$

The developers of the bag distance have shown that the property $\text{dist}_{\text{bag}}(s_1, s_2) \leq \text{dist}_{\text{levenshtein}}(s_1, s_2)$ always holds, and thus that $\text{sim}_{\text{bag}}(s_1, s_2) \geq \text{sim}_{\text{levenshtein}}(s_1, s_2)$. Because of this property, and because the computation complexity of the bag distance is $O(|s_1| + |s_2|)$, this distance can be used as an efficient filtering technique to remove string pairs before they are being compared using the computationally more complex Levenshtein edit distance. If one is, for example, only interested in pairs of strings that have a minimum similarity of θ, with $0 < \theta < 1$, then all string pairs s_1 and s_2 that have $\text{sim}_{\text{bag}}(s_1, s_2) < \theta$ can be removed after having been compared using the bag distance.

5.10.2 Compression Distance

The idea of using a compression function for similarity calculations has recently been proposed and investigated for application in areas such as clustering of biological sequences, optical character recognition, and music [77]. The *normalised compression distance* (NCD) has been defined as:

$$\text{dist}_{ncd}(s_1, s_2) = \frac{|C(s_1 s_2)| - \min(|C(s_1)|, |C(s_2)|)}{\max(|C(s_1)|, |C(s_2)|)}, \tag{5.26}$$

with C being a compression function (e.g. *Zlib* or *BZ2*), $| \cdot |$ the length of a compressed string, and $s_1 s_2$ the concatenation of the two input strings s_1 and s_2 [77]. The theoretical properties of the NCD have been investigated with regard to the performance of different compression algorithms [49]. These studies showed that, depending upon the type of data at hand, a careful consideration of the compression algorithm is required. The NCD works best on data with some structure. Note that the compression distance is not a proper mathematical distance metric as was discussed in Sect. 5.1 on p. 102. Specifically, in practice, due to the imperfections of the compressor C, the NCD can return a result in $0 \leq \text{dist}_{ncd}(s_1, s_2) \leq 1 + \varepsilon$, for some small ε [77]. The compression based similarity is then calculated as:

$$\text{sim}_{ncd}(s_1, s_2) = 1.0 - \text{dist}_{ncd}(s_1, s_2). \tag{5.27}$$

A recent study has shown that compression based similarity can be employed for data matching [166], and that it is suitable for situations where the data contain large amounts of noise in the form of typographical variations or errors. Compression based similarity is best applied on full records that contain several attribute values rather than on individual cleaned and standardised attributes. An additional aspect of this similarity function is that, besides the choice of compression function, it does not need any further parameters. This aspect makes compression distance an attractive technique for systems where automated data matching is required.

5.10.3 Editex

Editex is an approximate string comparison function that combines the phonetic information contained in two strings with edit distance based calculations [302]. Similar to Soundex (as was discussed in Sect. 4.3), the second and all following characters in the two strings s_1 and s_2 are mapped into a Soundex-like numerical code. As is done with edit distance, a dynamic programming approach is employed with edit cost 0 if two characters are the same, cost 1 if two characters are in the same phonetic group (i.e. have the same numerical code), and cost 2 if they are in different groups. Similar to Levenshtein edit distance, the time and space complexities of Editex are $O(|s_1| \times |s_2|)$ and $O(\min(|s_1|, |s_2|))$, respectively.

While Editex was originally developed within an information retrieval system [302], its properties of combining phonetic and edit distance based similarity calculations can make it attractive for data matching of attributes that contain personal names, because such names commonly contain phonetic variations. An experimental comparison by the developers of Editex using a database of 30,000 surnames has shown that it can outperform other string comparison functions including edit distance and q-gram based comparisons [302].

5.10.4 Syllable Alignment Distance

This technique, named *Syllable Alignment Pattern Searching (SAPS)* [125], also combines phonetic information with edit distance based calculations. The basic idea is to convert the strings to be compared into their corresponding sequences of syllables, and to calculate the number of edits that are required to convert one sequence of syllables into the other. The Phonix (see Sect. 4.3.3) phonetic transformation rules (without the final numerical encoding phase) are used to convert a string into an encoding according to how the string would be spoken (i.e. how it sounds). A set of linguistic rules are used to find the beginning of all syllables in a string, and using these rules the string is split into its syllables.

To find the syllable alignment distance, a dynamic programming approach similar to edit distance is then employed. Similar to the Smith-Waterman edit distance function discussed in Sect. 5.3.1, seven edit (or alignment) operations are considered with different edit scores, es:

- $es_m = 1$: Two characters (not syllable starts) are the same (they match).
- $es_d = -1$: Two characters (not syllable starts) are different.
- $es_{as} = -4$: The alignment of a character with a syllable start.
- $es_{ss} = 6$: Two syllable starts that are the same.
- $es_{ds} = -2$: Two syllable starts that are different.
- $es_{gc} = -1$: The alignment of a gap with a character (not a syllable start).
- $es_{gs} = -3$: The alignment of a gap with a syllable start.

From the resulting distance, a similarity can be calculated in the same three ways as for the Smith-Waterman edit distance described on p. 106.

The developers of SAPS [125] presented experiments indicated that this approximate string comparison function performs better than Editex, edit distance and Soundex, on the same database of names that was used for evaluating Editex [302].

5.11 String Comparison Examples

To illustrate how different approximate string comparison functions return different numerical similarity values when applied on the same strings, Figs. 5.4 and 5.5 show example results when comparing surnames and given names.

In Fig. 5.4, the similarity values calculated on selected individual name pairs are shown. They illustrate how the different approximate string comparison functions presented in this chapter return different similarity values for the same pair of name strings. As can be clearly seen, there are some large differences in the similarity values calculated on the same pair.

To get a more general picture of the similarity values calculated by the different string comparison functions, Fig. 5.5 shows the results of the comparison of the 5,000 most frequent given names found in an Australian telephone directory database. Each

String 1	String 2	Jaro	Winkler	Bigram	Trigram	PBigram	SkGram	LE-Dist	DLE-Dist	BagDist	Editex	ComZip	LCS2	LCS3	SW-Dist	SyllADist
shackleford	shackelford	0.970	0.982	0.750	0.692	0.750	0.821	0.818	0.909	**1.000**	0.806	0.684	0.818	0.818	0.691	*0.645*
dunningham	cunnigham	*0.896*	**0.896**	0.667	*0.522*	0.667	0.722	0.800	0.800	0.800	0.769	0.722	0.842	0.842	0.716	0.678
nichleson	nichulson	0.926	**0.956**	0.700	0.636	0.700	0.696	0.778	0.778	0.889	0.769	0.647	0.778	0.778	*0.622*	0.698
jones	johnson	0.790	**0.832**	0.429	*0.250*	0.286	0.355	0.429	0.429	0.571	0.500	0.533	0.333	0.333	0.333	0.500
massey	massie	0.889	**0.933**	0.571	*0.500*	0.571	0.645	0.667	0.667	0.833	0.714	0.714	0.667	0.667	0.733	0.774
abroms	abrams	0.889	0.922	0.714	0.625	0.714	0.677	0.833	0.833	0.833	0.889	0.571	0.833	*0.500*	0.900	**0.905**
hardin	martinez	0.722	0.722	0.250	*0.000*	0.250	0.306	0.500	0.500	0.500	0.542	0.500	0.571	0.286	0.629	0.078
itman	smith	0.467	0.467	0.167	0.000	0.167	0.115	0.000	0.000	**0.600**	0.133	**0.615**	0.400	0.400	0.400	*0.000*
jeraldine	geraldine	0.926	0.926	0.800	0.727	0.800	0.848	0.889	0.889	0.889	0.926	0.882	0.889	0.889	**0.933**	*0.714*
marhta	martha	0.944	0.961	0.571	*0.500*	0.571	0.710	0.667	0.833	**1.000**	0.765	0.571	0.500	0.500	0.500	0.615
michelle	michael	0.869	0.921	0.588	*0.421*	0.588	0.597	0.625	0.625	0.750	0.700	0.625	0.800	0.533	0.640	**0.930**
tanya	tonya	0.867	**0.880**	0.667	0.571	0.667	0.654	0.800	0.800	0.800	0.857	0.769	0.600	0.600	**0.880**	0.867
dwayne	duane	0.822	**0.840**	0.462	0.400	0.462	0.456	0.667	0.667	0.667	0.688	0.571	*0.364*	*0.364*	*0.364*	0.759
sean	susan	0.783	**0.805**	0.545	*0.462*	0.545	0.468	0.600	0.600	0.600	0.667	0.615	0.444	0.444	0.489	0.529
jon	john	0.917	**0.933**	0.667	0.545	0.667	0.595	0.750	0.750	0.750	0.818	0.667	0.571	0.571	0.571	0.889
brookhaven	brrokhaven	0.933	0.947	0.909	*0.750*	0.909	0.843	0.900	0.900	0.900	**1.000**	0.778	0.900	0.700	0.800	0.769
higbee	highee	0.889	**0.922**	0.714	0.625	0.714	0.677	0.833	0.833	0.833	0.800	0.571	0.833	0.500	0.667	*0.312*
cunningham	cunnigham	0.967	**0.980**	0.857	0.783	0.857	0.866	0.900	0.900	0.900	0.885	*0.722*	0.947	0.947	0.821	0.949
campell	campbell	0.958	**0.975**	0.824	0.737	0.824	0.779	0.875	0.875	0.875	0.900	0.688	0.933	0.933	0.773	*0.578*
galloway	calloway	0.917	**0.917**	0.778	0.700	0.778	0.829	0.875	0.875	0.875	0.895	0.875	0.875	0.875	0.875	*0.652*
michele	michelle	0.958	0.975	0.941	*0.842*	0.941	0.857	0.875	0.875	0.875	**1.000**	0.750	0.800	0.800	0.800	**1.000**
jonathon	jonathan	0.917	**0.950**	0.778	0.700	0.778	0.780	0.875	0.875	0.875	0.913	0.750	0.750	*0.750*	0.925	0.929
dickson	dixon	0.790	**0.832**	0.571	0.500	0.571	0.387	0.571	0.571	0.571	0.684	0.533	0.667	*0.333*	*0.333*	0.837
gail	gayle	0.783	**0.827**	0.364	*0.308*	0.364	0.426	0.600	0.600	0.600	0.643	0.615	0.444	0.444	0.444	0.857
sydney	sydeny	0.944	0.961	0.571	*0.500*	0.571	0.710	0.667	0.833	**1.000**	0.667	0.571	0.500	0.500	0.500	0.486
tsetung	zedong	**0.643**	**0.643**	0.267	*0.235*	0.267	0.239	0.429	0.429	0.429	0.571	0.533	0.308	0.308	0.585	*0.143*

Fig. 5.4 Example string pairs and the similarities calculated on them. 'PBigram' stands for positional bigrams, 'SkGram' for skip-grams, 'LE-Dist' for the Levenshtein edit distance, 'DLE-Dist' for the Damerau-Levenshtein edit distance, 'ComZip' for compression based similarity using the ZLib compressor, 'LCS2' and 'LCS3' for the longest common substring comparison with the minimum length set to 2 and 3, respectively, 'SW-Dist' denotes the Smith-Waterman edit distance and 'SyllADist' the syllable alignment distance. In each row, the largest similarity value is shown in boldface and the lowest in italics

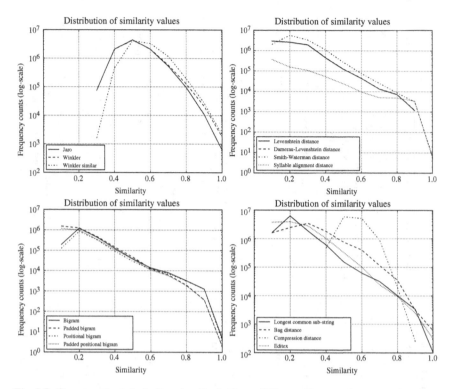

Fig. 5.5 The normalised similarity value distributions of 15 approximate string comparison functions based on comparing 5,000 given name values with each other. The Dice coefficient was used in all comparison functions that require the conversion of a set size into a similarity

given name in this database was compared with all others, and all similarity values sim > 0.0 were recorded and rounded to one digit to smooth the result curves. Because names in this database mostly consist of one short word only, results for the Winkler long adjustment, and the multi-word comparison functions (Monge-Elkan, SoftTFIDF and extended Jaccard) are not shown.

As can be seen, the different comparison functions have very different characteristics in the average as well as the spread of similarity values they return. The Jaro, Winkler and the compression based comparison functions have the highest average similarity values, while the edit distance, the q-gram based and the LCS comparison functions return much lower average similarities.

An important aspect when using different similarity functions together with a threshold-based classifier (to be discussed in the following chapter) is that any minimum similarity threshold used to classify candidate record pairs into matches and non-matches needs to be adjusted, when different approximate string comparison functions are employed.

Fig. 5.6 Illustration of the linear extrapolation of the numerical absolute difference similarity given in Eq. 5.28

5.12 Numerical Comparison

For certain applications, the data used in data matching not only contain string values, but also attributes that contain numerical information. A commonly occurring example is financial data such as salaries, savings, expenses or taxation amounts.

It is assumed two numerical values, n_1 and n_2, one each from the two records in a candidate record pair, are compared with each other. If both values are the same, $n_1 = n_2$, their similarity will be $sim(n_1, n_2) = 1.0$. Similar to approximate string comparison functions, to allow for variations and errors in numerical data it must be possible to calculate approximate similarities between numerical values. This can be achieved with two different approaches.

In the first approach, a maximum absolute difference, d_{max} ($d_{max} > 0$), is tolerated between the two values n_1 and n_2, independently of their actual values. For this approach, the similarity between n_1 and n_2 is calculated as:

$$\text{sim}_{\text{num\_abs}} = \begin{cases} 1.0 - \left(\frac{|n_1 - n_2|}{d_{max}} \right) & \text{if } |n_1 - n_2| < d_{max}, \\ 0.0 & \text{else.} \end{cases} \tag{5.28}$$

For a pair of values that have an absolute difference smaller than what is tolerated, i.e. $|n_1 - n_2| < d_{max}$, a linear extrapolation between exact similarity (1.0) and total dissimilarity (0.0) is calculated with Eq. (5.28), as illustrated in Fig. 5.6.

For example, assume the maximum absolute difference in two salary values that is to be tolerated is set to $d_{max} = \$1,000$. If $n_1 = \$2,000$ and $n_2 = \$2,500$ then $\text{sim}_{\text{num\_abs}}(\$2,000, \$2,500) = 1.0 - \frac{\$500}{\$1,000} = 1.0 - 0.5 = 0.5$. On the other hand, if $n_1 = \$450,000$ and $n_2 = \$450,250$ then $\text{sim}_{\text{num\_abs}}(\$450,000, \$450,250) = 1.0 - \frac{\$250}{\$1,000} = 1.0 - 0.25 = 0.75$.

These two examples highlight an issue of this absolute difference approach. The approximate similarity value calculated is independent of the numerical values compared. In certain situations, rather than having an absolute difference tolerated, a relative difference could be more appropriate. Such a relative numerical similarity value can be calculated based on the percentage difference, pc, between two values, n_1 and n_2, as:

$$pc = \frac{|n_1 - n_2|}{\max(|n_1|, |n_2|)} \cdot 100. \tag{5.29}$$

Similar to the absolute numerical difference, d_{max}, a maximum percentage difference, pc_{max} $(0 < pc_{max} < 100)$ is used to calculate the approximate similarity value:

$$
\text{sim}_{\text{num\_perc}} = \begin{cases} 1.0 - \left(\frac{pc}{pc_{max}}\right) & \text{if } pc < pc_{max}, \\ 0.0 & \text{else.} \end{cases} \tag{5.30}
$$

Using the same pair of example values from above, and setting $pc_{max} = 33\,\%$, then for the first pair of values, $n_1 = \$2,000$ and $n_2 = \$2,500$, the numerical percentage difference is calculated as $pc = \frac{|\$2,000-\$2,500|}{\max(\$2,000,\$2,500)} \cdot 100 = \frac{\$500}{\$2,500} \cdot 100 = \frac{1}{5} \cdot 100 = 20\,\%$. From this, the numerical percentage similarity is then calculated as $\text{sim}_{\text{num\_perc}}(\$2,000, \$2,250) = 1.0 - \frac{20}{33} = 1.0 - 0.606 = 0.394$. On the other hand, following similar calculations for the second pair gives $pc = \frac{\$250}{\$450,250} \cdot 100 = 0.0555\,\%$, and thus $\text{sim}_{\text{num\_perc}}(\$450,000, \$450,250) = 1.0 - \frac{0.0555}{33} = 1.0 - 0.0017 = 0.9983$.

Which of these two approaches to use and what maximum absolute or percentage differences to tolerate depend upon the data at hand and how much variation can be tolerated for a given data matching application.

5.13 Date, Age and Time Comparison

Given data matching is commonly based on the comparison of personal information, the ability to compare dates, times and age values are of importance. Dates can be seen as a special case of numerical data.

For dates, the commonly used approach is to calculate the difference between two dates in days, and then to employ an absolute day difference comparison based on the numerical absolute difference approach described in Eq. (5.28) above. For example, if an absolute difference of $d_{max} = 30$ days (around one month) is tolerated, and the two dates to be compared are $d_1 = [11, 03, 2010]$ and $d_2 = [29, 03, 2010]$, with a difference of 18 days, then the approximate similarity would be $\text{sim}_{\text{day\_abs}}(d_1, d_2) = 1.0 - \frac{18}{30} = 1.0 - 0.6 = 0.4$.

Compared to generic numerical data, there are two special cases of errors that frequently occur with dates, and that therefore need to be considered.

- If the day and month values are swapped in the date values in two records, but the year value is the same, then an approximate similarity value should be returned, for example, $\text{sim}_{\text{day\_abs}} = 0.5$, even if the two dates have a day difference outside of the maximum d_{max}. This situation can occur when the day and month values are entered in the wrong order, or mistakenly entered in the US format [MM, DD,YYYY] rather than the format used in many other countries, [DD, MM,YYYY], or the other way around. For example, [12,10,2010] and [10,12,2010] could be two such dates where day and month values have been swapped.

The problem mainly occurs if both day and month values are between 1 and 12, because in these situations it will not be possible to automatically detect an error. If a month value entered is above 12, then it likely refers to the day value and the person doing the data entry can either be alerted about this potential error, or the day and month values can be swapped automatically.

- If the difference between two dates is only in the month value but both year and day values are the same, and the day difference is larger than the maximum tolerated value, d_{max}, then an approximate similarity value should be calculated that is larger than the similarity calculated when day and month values are swapped. The reason for this is that two out of three date elements are exact matches, such as in the example [12,01,2010] and [12,07,2010]. As a full date consists of eight digits, and only one or two of them are different in this situation, an approximate similarity of at least $sim_{day\_abs} = 0.75$ can be used.

An alternative way to compare date values that, for example, refer to dates of birth, is to convert date values into age values which can be compared with a certain percentage difference that is tolerated, as was discussed in the previous section. Age values need to be calculated according to a certain fix date, which can either be the date of when two databases are being matched, or any other specific date appropriate to the matching problem at hand.

Ideally, dates are converted into a number of days or years, such as the number of days or years since birth. Assuming d_1 and d_2 are the two day values that are being compared, the age percentage difference, apc, is calculated similar to Eq. (5.29) as:

$$apc = \frac{|d_1 - d_2|}{\max(d_1, d_2)} \cdot 100. \qquad (5.31)$$

Based on this age percentage difference, the age percentage similarity is then calculated as:

$$sim_{age\_perc} = \begin{cases} 1.0 - \left(\dfrac{apc}{apc_{max}}\right) & \text{if } apc < apc_{max}, \\ 0.0 & \text{else}, \end{cases} \qquad (5.32)$$

with apc_{max} ($0.0 < apc_{max} < 100$) the maximum age percentage difference that is tolerated.

There are also data matching applications where time values need to be compared. This can be accomplished similar to dates using either an approach based on tolerating absolute time differences (following Eq. (5.28)), or percentage differences relative to a certain fix time (following Eq. (5.32)). Time differences used in the comparison calculations can be calculated in hours, minutes or seconds, depending upon the resolution required by a data matching application.

5.14 Geographical Distance Comparison

Increasingly, geographical information is becoming available in many applications, for example, in the form of geographic locations (longitude and latitude) for addresses.

Rather than using only the similarity between address components (consisting of strings), an alternative way to calculate the similarity between two addresses is to calculate their geographical distance, and use this numerical value for a numerical comparison as discussed in Sect. 5.12. Geographical distances are measured along the surface of the Earth in kilometres or miles. A large body of work is available in the geographical literature on how to calculate geographic distances based on various projection methods. Freely available software libraries such as Geographiclib[3] and online tools[4] can help implementing such comparison functions.

The topic of geocode matching, the matching of addresses to their geographic locations, will be further discussed in Sect. 9.1. Within the context of comparing addresses according to their geographical locations, it is important to note that due to data quality problems it is often not possible to obtain the exact and accurate location of an address. If address details are missing (for example, no street number is given, or only postcodes or zipcodes are available in a database), or if they contain erroneous values, then the location might only be accurate at the level of a region, such as a street or suburb. Regions can be represented through their bounding boxes, the smallest rectangle that encloses a region, or by the location of the centre of a region and its radius. The issues of geocode matching accuracy will be covered further in Sect. 9.1.

5.15 Comparing Complex Data

While thus far only the comparisons of simple, atomic types of data have been discussed, increasingly databases contain more complex objects such as XML documents, or multimedia data (images, audio and video). When comparing such complex data, an appropriate object description needs to be selected [195].

For XML data, one possible approach is to map XML documents into a relational database table, where each document is split into several tuples, each consisting of a value and a name type. The value is the actual data taken from an XML element, while the name type identifies the type of the data [270]. Examples of such tuples are (Miller', 'Name/Surname'), ('42', 'Address/StreetNumber'), or ('Main', 'Address/StreetName'). The values in the tuples that have the same name type are then compared pair-wise using one of the approximate string comparison functions described in this chapter. For each pair of XML records an overall similarity can be calculated based on the common or similar values they have in the same type.

---

[3] See: http://geographiclib.sourceforge.net/

[4] See: http://geographiclib.sourceforge.net/cgi-bin/Geod

RecID	GivenName	Surname	StrNum	StrName	Suburb	SimSum
a1	john	smith	18	miller st	dickson	
b1	jonny	smyth	73	miller st	dixon	
	0.6	0.8	0.0	1.0	0.6	3.0

RecID	GivenName	Surname	StrNum	StrName	Suburb	SimSum
a2	mary	harris	42	swamp rd	sydney	
b2	mandy	garrett	42	smither rd	sydenham	
	0.6	0.4	1.0	0.4	0.6	3.0

Fig. 5.7 Example comparison vectors and their summed similarities (SimSum), illustrating the loss of detailed information that occurs when only summed similarity values, rather than full comparison vectors, are used for classification. The first pair of records likely corresponds to the same individual, as there are only small differences in the two name fields and the suburb field, and the different digits in the street number could be due to optical character recognition mistakes. The second pair, while having the same overall similarity value (SimSum), is unlikely to correspond to the same person. All similarities in this example were calculated using the Levenshtein edit distance and were rounded to 1 digit

Several other techniques have been developed to compare XML documents, the interested reader can refer to the book by Naumann and Herschel [195] for details.

For calculating the similarity between multimedia data, various techniques are available. They are based on the concept of extracting features from a media file, for example, a colour histogram of an image, and to store such features into feature vectors. Similarities are then calculated based on comparing these feature vectors [135].

5.16 Record Comparison

For each candidate record pair that is compared, generally several attributes are compared in detail, as illustrated in Fig. 2.6 on p. 31. A vector of numerical similarity values, commonly called a 'comparison vector', is generated for each record pair. These comparison vectors are the basis of most classification techniques for data matching, as will be discussed in the following chapter.

While with traditional data matching approaches all values of a comparison vector are summed into a single similarity for a candidate record pair (denoted with 'SimSum' in Figs. 2.6 and 5.7), more advanced classification techniques can provide a further weighting scheme to the basic normalised similarity values, giving larger weights to attributes that contain information that is more distinctive. Surname values, for example, are much better indicators if two records refer to the same or different individuals compared to a gender value. The summation of all similarity values in a comparison vector into a single similarity value results in a severe loss of information, as the examples in Fig. 5.7 show.

5.17 Practical Considerations and Research Issues

One question that is sometimes difficult to answer in practical data matching and deduplication projects is which of the many available comparison functions should be used. While there is a limited number of comparison functions available for numerical attributes, or for those that contain date, age, time or location values, for attributes that contain strings the choice of comparison function can be rather large (even though some commercial data matching systems might only contain a small number of approximate string comparison functions).

A starting point is to explore the content of the string attributes that will be used in the comparison. Do they only contain short strings made of one word (or token) only, or do they contain longer strings consisting of several words? In the latter case, a proper segmentation of the input data might be needed first (as was discussed in Chap. 3), especially if the values in a single attribute consist of several pieces of different information (for example, if full names or full addresses are contained in one long string attribute). Alternatively, one of the comparison functions (such as Monge-Elkan, SoftTFIDF and extended Jaccard) that can handle multiple words (tokens) in the input strings can be employed.

For attributes that contain name values (such as given names, surnames, street names, location/town names or state/territory names), it has been shown that the Jaro-Winkler approximate string comparison function (discussed in Sect. 5.5) performs better compared with other comparison functions [294]. For attributes that contain shorter non-name string values, edit distance or q-gram based techniques can be employed, with edit distance based techniques having a larger computation complexity than q-gram based techniques. For longer strings, edit distance based techniques will likely become too slow in practice. For attributes where string values contain several words (or more generally several tokens), one of the functions presented in this chapter that are specialised for multi-word comparisons will be the best choice.

If the available data contain the true match status of record pairs, for example, from a matching or deduplication exercise on an earlier version of the same database(s), or from a manual evaluation of selected record pairs, then different comparison functions can be applied in a series of test matching exercises. Such an evaluation needs to be conducted on the set of record pairs where the match status is available. The set of comparison functions that results in the best separation of matches from non-matches can then be used for matching the full databases.

Research in the area of comparison functions for data matching can for example, be aimed at tackling the problem of multi-lingual name values and how they are best compared, such that variations of the same name result in high similarity values and variations of different names result in low similarity values. As most data are changing over time, developing comparison functions that adapt themselves to changing data is another avenue for research. Such adaptive functions will likely need to have access to training data in the form of true matching and true non-matching string pairs [35].

With regard to computation complexity, developing comparison functions that have a complexity that is linear in the length of the two compared strings would lead to a much improved scalability of data matching systems. Many of the comparison functions presented in this chapter have a computation complexity that is quadratic in the length of the two strings. Approximate or heuristic approaches can be investigated to reduce this computation complexity.

A further investigation of heuristic adjustments, such as the Winkler adjustment to the basic Jaro comparison function, based on studies of real-world data and their errors and variations, could also lead to improved data matching quality. A related topic is the investigation of how different approximate string comparison functions can handle data with different types of errors (such as optical character recognition, phonetic or manual key-board typing errors), and how string comparison functions can be adapted to such different types of data characteristics.

Finally, the overall question of which comparison function is best suited for what type of data could be investigated through large-scale experimental studies using a variety of data sets with different content (such as personal names and addresses, business names, consumer product names, details of scientific or technical publications and so on), different error characteristics and values from different languages.

5.18 Further Reading

Many surveys of approximate string comparison functions have been published over the past three decades [57, 119, 133, 152, 172, 175, 195, 196, 243, 294]. There are also several open source libraries available that implement a variety of comparison techniques: SecondString[5] [84], FEBRL[6] [62], and SimMetrics[7] are some of the most popular ones. Many modern database systems also contain extensions that allow approximate similarity calculations of strings.

A detailed description of the Jaro and Winkler approximate string comparison functions is provided by Herzog, Scheuren and Winkler [143]. For the comparison functions of other data types, such as numerical, date, age, time and geographical distance, not many technical publications are available. However, such comparison functions have been implemented and are used successfully in many (commercial) data matching systems.

---

[5] See: http://sourceforge.net/projects/secondstring/

[6] See: http://sourceforge.net/projects/febrl/

[7] See: http://sourceforge.net/projects/simmetrics/

Chapter 6
Classification

6.1 Overview

The classification of the candidate record pairs that were generated in the indexing step (Chap. 4) and compared in detail in the comparison step (Chap. 5) is primarily based on the similarity values in the comparison vectors of these record pairs, as illustrated in Fig. 6.1. The general idea is that the more similar two records are, the more likely they refer to the same real-world entity.

A classification approach can either be unsupervised or supervised. Unsupervised approaches classify pairs or groups of records based on similarities between them without having access to any information about the characteristics of true matching and true non-matching record pairs. Supervised approaches, on the other hand, require training data that are known true matches and true non-matches. Specifically, a set of comparison vectors is required, each of which has a match status (match or non-match) attached, to enable training of a supervised classifier. These comparison vectors need to be generated using the same comparison functions as the ones that will be used when pairs of records with unknown match status are compared. Obtaining or generating such training data, that need to be of high quality and cover a large variety of the possible similarity value combinations that can occur in comparison vectors, can be difficult, as will be discussed further in Sect. 7.1 in the context of evaluating the outcomes of data matching classification.

In certain matching situations, additional to the similarities between records, relational information between records might be available. Examples of such relational information include the lists of co-authors of scientific publications, or people who share the same land-line telephone number. Several recently developed classification techniques for data matching take such relational links or connections into account when building a clustering or graph-based classification model [31, 142, 155, 272], as will be discussed in Sects. 6.9 and 6.10. An alternative approach to model such relational information is to include it as exact similarities into the comparison vectors for candidate record pairs. As was discussed in Sect. 5.2, an exact comparison would for example, result in a normalised similarity value of 1.0 if two records have the

P. Christen, *Data Matching*, Data-Centric Systems and Applications,
DOI: 10.1007/978-3-642-31164-2_6, © Springer-Verlag Berlin Heidelberg 2012

RecID	GivenName	Surname	StrNum	StrName	Suburb	BDay	BMonth	BYear	SimSum
a1	john	smith	18	miller st	dickson	12	11	1970	
b1	jonny	smyth	73	miller st	dixon	11	12	1970	
	0.6	0.8	0.0	1.0	0.6	0.5	0.5	1.0	5.0
a2	mary	harris	42	swamp rd	sydney	21	04	1918	
b2	mandy	garrett	42	smither pl	sydenham	27	04	1979	
	0.6	0.4	1.0	0.4	0.6	0.5	1.0	0.5	5.0

Fig. 6.1 Example candidate record pairs and their comparison vectors as calculated in the comparison step (Chap. 5), and their summed similarity values (SimSum)

same telephone number, or if two publications have been written by the same two co-authors.

When the classification into matches and non-matches is conducted independently for individual candidate record pairs rather than in a collective fashion [31], then several issues need to be considered. First, classifying each record pair independently can lead to sub-optimal match decisions. For example, assume two databases are matched, with the restriction that a record from the first database can only match to a maximum of one record in the second database (a topic that will be discussed further in Sect. 6.11). Now assume that record 'a1' (from the first database) is classified to match with record 'b4' (from the second database). Later on in the classification process, record 'a9' and 'b4' are also found to be a match, for example, if their similarity is higher than the similarity between 'a1' and 'b4'. The first classified match between 'a1' and 'b4' is therefore not the best match. Such sub-optimal decisions can occur if the classification is conducted in a greedy fashion where, once matched, a pair of records is not reconsidered for other matches.

A second issue is that independent match decisions can lead to contradictions, an issue known as transitive closure [190]. Assume in the deduplication of a database two record pairs, 'a1' and 'a2', and 'a1' and 'a3', have both been classified as matches, but the pair 'a2' and 'a3' has been classified as a non-match. This contradicts the assumption that when a group of three or more records are classified as being duplicates of each other, then none of the individual pairs in that group of records should be classified as a non-duplicate. How to handle this situation will be further discussed in Sect. 6.8.

After the classification step, depending upon the data matching or deduplication situation, records that were matched in certain applications need to be merged into compound new records [26]. This step itself is challenging, as it requires decisions to be made about how to merge individual attribute values that potentially can contradict with each other. A set of recently developed techniques to accomplish this will be presented in Sect. 6.12.

6.2 Threshold-Based Classification

The simplest way to classify candidate record pairs into the two classes of matches and non-matches (and possibly the third class of potential matches) is to sum the similarity values in their comparison vectors into a single total similarity value (called 'SimSum' in Fig. 6.1), and to then apply a similarity threshold (or two in the case where potential matches are considered) to decide into which class a candidate record pair belongs. Figure 6.2 shows an example histogram of the summed comparison vectors obtained from a deduplication of a real health data set that contained 175,211 records.

With two classes (matches and non-matches), a single classification threshold, t, is needed for classifying a record pair (r_i, r_j):

$$SimSum[r_i, r_j] \geq t \Rightarrow [r_i, r_j] \rightarrow \text{Match},$$
$$SimSum[r_i, r_j] < t \Rightarrow [r_i, r_j] \rightarrow \text{Non-Match}. \tag{6.1}$$

With three classes (matches, non-matches and potential matches), two classification thresholds, t_l (lower) and t_u (upper), are needed, and a record pair (r_i, r_j) is classified according to:

$$SimSum[r_i, r_j] \geq t_u \Rightarrow [r_i, r_j] \rightarrow \text{Match},$$
$$t_l < SimSum[r_i, r_j] < t_u \Rightarrow [r_i, r_j] \rightarrow \text{Potential Match}, \tag{6.2}$$
$$SimSum[r_i, r_j] \leq t_l \Rightarrow [r_i, r_j] \rightarrow \text{Non-Match}.$$

Selecting a threshold (or thresholds) that results in high matching quality can either be done manually or (if training data are available) be learned in such a way that either the number of false matches or the number of false non-matches is minimised, or that the sum of both false matches and false non-matches is minimised. When candidate record pairs are also classified into potential matches, there is a trade-off between the quality of the classified matches and non-matches and the amount of manual clerical review of the potential matches that needs to be conducted. This topic will be discussed in more detail in Sect. 7.4.

If $t_u = t_l = t$ then the class of potential matches disappears and the classification follows Eq. 6.1. Increasing the upper threshold t_u and lowering the lower threshold t_l will result in less false matches and non-matches, but lead to a larger number of potential matches that need to be inspected and classified manually. On the other hand, lowering t_u and increasing t_l results in a smaller number of potential matches but likely also in an increased number of false matches and non-matches. This issue will be discussed further in Sect. 7.4.

The simple threshold-based classification has two major drawbacks. The first is that, assuming all similarity values are normalised between 0 and 1, all attribute similarities contribute in the same way towards the final summed similarity value. The importance of different attributes, as well as their discriminative power with

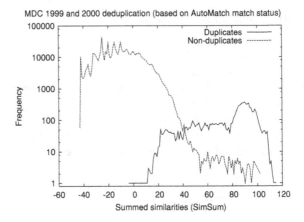

Fig. 6.2 Example histogram of the summed similarity values of a deduplication of a real health data set using comparisons on twelve attributes, with different weights assigned to the various comparisons. The true match status of this data set was determined earlier using the commercial data matching software AutoMatch [251], while the comparison vectors for this deduplication were generated using the FEBRL system [62]. A detailed description of the settings that were used for this deduplication exercise is provided in [71], from where this figure has been adapted

regard to distinguishing matches from non-matches, is not considered by such a simple summation approach. This drawback can be overcome by summing similarity values that are weighted, with different attributes given different weights according to their importance or discriminative power. The weighed sum is calculated by first multiplying the similarity value calculated on a certain attribute with the weight value for this attribute prior to the summation.

For example, assuming the following weights (empirically determined) are assigned to the attributes of Fig. 6.1: $w_{GivenName} = 2$, $w_{Surname} = 3$, $w_{StrNum} = 1$, $w_{StrName} = 3$, $w_{Suburb} = 2$, $w_{BDay} = 2$, $w_{BMonth} = 1$, and $w_{BYear} = 2$. The weighted summed similarities for the two given record pairs then become: $SimSum[a1, b1] = 2 \times 0.6 + 3 \times 0.8 + 3 \times 1.0 + 2 \times 0.6 + 2 \times 0.5 + 1 \times 0.5 + 2 \times 1.0 = 11.3$, and $SimSum[a2, b2] = 2 \times 0.6 + 3 \times 0.4 + 1 \times 1.0 + 3 \times 0.4 + 2 \times 0.6 + 2 \times 0.5 + 1 \times 1.0 + 2 \times 0.5 = 8.8$. The question of how to choose good weight values for the different attributes will be discussed in the next section.

The second drawback of summing similarities is that the detailed information contained in the individual similarity values is lost in the summation step (with both an unweighted or weighted approach). As the two example candidate record pairs in Fig. 6.1 show, despite having different similarity values both result in an unweighted sum of SimSum = 5.0. This is even though record pair (a2, b2) is unlikely to refer to the same entity given the differences in the attribute values of these two records. Giving different weights to attributes can to some degree overcome this drawback and lead to better classification results. More sophisticated classifiers, presented later in this chapter, are utilising the individual similarity values. This generally leads to improved matching quality compared to the simple threshold-based approach using summed similarities only.

6.3 Probabilistic Classification

This traditional classification approach to data matching, proposed in 1969 by Ivan Fellegi and Alan Sunter in their seminal paper [108], is commonly known as 'probabilistic record linkage'. Many data matching and deduplication systems that have been developed over the past four decades are based on the approach described in this paper.

The basic ideas of probabilistic record linkage were introduced by Newcombe et al. [198] in 1959 and detailed further by Newcombe and Kennedy in 1962 [197]. They recognised that in the absence of unique entity identifiers the attributes available in common in two databases (such as the names, addresses, or dates of birth of patients or customers) need to be used to match records. As the values in such attributes can be wrong, missing, or out of data, and because the number of values and their distributions can differ between attributes, different weights should be assigned to different attributes when they are used to calculate the similarities between records (as was discussed in the previous section). Newcombe and Kennedy also recognised that such weights should not only depend upon the general characteristics of attributes, but also on the actual attribute values in a certain candidate record pair. For example, if two records have a surname value 'Smith' then the weight given for this agreement of values should be smaller than the weight given to two records that both have a surname value 'Dijkstra', assuming the number of people with surname 'Dijkstra' is much smaller than the number of people with surname 'Smith' in the databases that are matched. This is because the likelihood that two randomly picked records have a surname value 'Smith' is much higher than the likelihood that they have a surname value 'Dijkstra'.

Fellegi and Sunter formalised these ideas and they developed a theory for record linkage that allows the calculation of weights for agreeing and disagreeing pairs of attribute values, which leads to an optimal decision making when record pairs are classified [108]. Probabilistic record linkage considers two databases (or files), **A** and **B**, and record pairs in the product space $\mathbf{A} \times \mathbf{B}$ that are to be classified into three classes: Matches (links), non-matches (non-links) and potential matches (potential links) [108, 143]. Record pairs classified as potential matches need to be manually assessed and classified in a clerical review process, as will be described in Sect. 7.4. Each record pair in $\mathbf{A} \times \mathbf{B}$ is assumed to correspond to either a true match or a true non-match. The space $\mathbf{A} \times \mathbf{B}$ is therefore partitioned into the set M of true matches and the set U of true non-matches. Formally,

$$\mathbf{A} \times \mathbf{B} = \{(a, b); \ a \in \mathbf{A}, b \in \mathbf{B}\} \tag{6.3}$$

consists of the two disjoint sets

$$M = \{(a, b); a = b, a \in \mathbf{A}, b \in \mathbf{B}\} \tag{6.4}$$

of true matches (also called the matched set), where both records a and b refer to the same real-world entity, and

$$U = \{(a, b); a \neq b, a \in \mathbf{A}, b \in \mathbf{B}\} \tag{6.5}$$

of true non-matches (also called the unmatched set), where the two records a and b refer to two different real-world entities.

The assumption is that records in $\mathbf{A}$ and $\mathbf{B}$ were generated based on one process for each of the two databases. Each record is assumed to refer to an individual in a population, such as a patient, customer, citizen and so on. The records in the two databases are drawn from two populations that have some overlap. For each member of the two populations, it is assumed that a record was generated with certain characteristics (such as certain name values, a certain date of birth and so on). The record generation process also led to errors and missing values in the records with certain distributions. As a result, it is possible that unmatched entities in $\mathbf{A}$ and $\mathbf{B}$ can be represented by two records that both have the same values in all their attributes. On the other hand, two records in $\mathbf{A}$ and $\mathbf{B}$ that refer to the same entity can have different values in some of their attribute.

When record pairs are compared (as was discussed in Chap. 5), a comparison vector, γ, is generated for each record pair. In the basic formulation of probabilistic record linkage, only binary comparisons are considered (with similarity value 1 when two attribute values are the same and 0 otherwise) [108]. Therefore, each γ corresponds to an agreement pattern in a comparison space, Γ. If each record pair was compared using K comparison functions, then each γ consists of a vector of K agreement or disagreement values. In total, assuming binary comparisons (i.e. exact matching) only, there will be 2^K different possible patterns.

For a given candidate record pair, r, probabilistic record linkage classification considers ratios of conditional probabilities, $P(\cdot|\cdot)$, of the form

$$R = \frac{P(\gamma \in \Gamma | r \in M)}{P(\gamma \in \Gamma \mid r \in U)} \tag{6.6}$$

where γ is an arbitrary agreement pattern in a comparison space Γ. Fellegi and Sunter [108] then propose the following decision rule:

$$R \geq t_u \Rightarrow r \rightarrow \text{Match,}$$
$$t_l < R < t_u \Rightarrow r \rightarrow \text{Potential Match,} \tag{6.7}$$
$$R \leq t_l \Rightarrow r \rightarrow \text{Non-Match.}$$

The two cutoff thresholds t_l and t_u are determined by a priori error bounds on false matches and false non-matches [108, 143]. It is easy to see that the three rules in Eq. 6.7 make intuitive sense. If γ for a certain candidate record pair r mostly consists of agreements, then the ratio R in Eq. 6.6 would be large, because it is more likely that $r \in M$ rather than $r \in U$, and the pair is more likely designated as a match. On the other hand, for a γ that primarily consists of disagreements the ratio R would be small, because it is more likely that $r \in U$ rather than $r \in M$, and thus the pair will be designated as a non-match.

Fellegi and Sunter showed that with fixed bounds on the errors in the match and non-match regions of R the decision rule in Eq. 6.7 is optimal, in that the middle region of potential matches is minimised [108].

Calculating the conditional probabilities in Eq. 6.6 is a crucial aspect of the probabilistic record linkage approach. It is commonly assumed that these probabilities are conditionally independent for the different attributes that are used in the comparison step to calculate the agreement patterns γ. Under this assumption, an individual agreement weight, w_i, $1 \leq i \leq K$ can be calculated for each attribute (or field) i based on the m- and u-probabilities

$$m_i = P([a_i = b_i, a \in \mathbf{A}, b \in \mathbf{B}] \mid r \in M), \tag{6.8}$$

and

$$u_i = P([a_i = b_i, a \in \mathbf{A}, b \in \mathbf{B}] \mid r \in U), \tag{6.9}$$

where a_i and b_i are the values in attribute i that are being compared. Equation 6.8 is the probability that two records have the same value in attribute i given the pair is a true match (i.e. both records refer to the same entity). On the other hand, Eq. 6.9 is the probability that two records have the same value in attribute i given the pair is a true non-match (i.e. the two records refer to different entities).

The probabilities in Eqs. 6.8 and 6.9 are called the m- and u-probabilities, respectively, and they are also known as the matching parameters [143]. Based on these two probabilities, the individual weight w_i for attribute i is calculated as:

$$w_i = \begin{cases} log_2(\frac{m_i}{u_i}) & \text{if } a_i = b_i, \\ log_2(\frac{(1-m_i)}{(1-u_i)}) & \text{if } a_i \neq b_i. \end{cases} \tag{6.10}$$

To make a simple example, assume the two databases $\mathbf{A}$ and $\mathbf{B}$ contain an attribute 'MonthOfBirth' (MoB) with twelve possible values 'January' to 'December'. Assume also that it is known that in both $\mathbf{A}$ and $\mathbf{B}$ this attribute contains 3% errors, i.e. 3% of all month of birth values have been recorded wrongly. The likelihood that two records, $a \in \mathbf{A}$ and $b \in \mathbf{B}$, that are known to refer to the same entity $((a, b) \in M)$ have the same month of birth value is 97%. Therefore, $m_{MoB} = 0.97$. The likelihood that two records referring to the same entity have a different month of birth is $(1 - m_{MoB}) = 0.03$ (3%). For two records $a \in \mathbf{A}$ and $b \in \mathbf{B}$ that are known to refer to two different entities $((a, b) \in U)$, the likelihood that their month of birth value is the same is $1/12 = 0.083$, because there is a $1/12$ (8.3%) chance that two randomly picked individuals in a population have the same month of birth. Therefore, $u_{MoB} = 0.083$. Conversely, the likelihood that two randomly picked records that refer to two different entities have a different month of birth is $11/12 = 0.917 = (1 - u_{MoB})$ (91.7%). Using Eq. 6.10, if two records have the same value in the 'MonthOfBirth' attribute, then the corresponding weight (called match or agreement weight) is calculated as $w_{MoB} = log_2(0.97/0.083) = 3.54$, while if two records have different month of birth values then the weight (called non-match or disagreement weight) is calculated as $w_{MoB} = log_2(0.03/0.917) = -4.92$.

Assuming conditional independence, the overall weight for a record pair r can be calculated by summing the weights w_i over the K attribute match/non-match weights:

$$log_2(R) = \sum_{i}^{K} w_i. \tag{6.11}$$

Figure 6.2 shows an example histogram of such summed weights for the deduplication of a real health data set. As can be seen, the number of non-matches (non duplicates) is much larger than the number of matches (duplicates), as would be expected.

In real-world data, it is likely that there are some dependencies between attributes. For example, records that have the same post- or zipcode with a high likelihood will also have the same locality (suburb or town) name, because in many countries most post- or zipcodes are contained within a certain locality. Records that have the same post- or zipcode then potentially also more likely have the same street name. However, despite most real-world data violating the conditional independence assumption, practical data matching projects have shown that good matching quality can still be achieved under this assumption [143].

One of the difficulties with probabilistic record linkage is the accurate calculation or estimation of the error rates required in Eqs. 6.8 and 6.9. Sometimes these probabilities are known from the manual assessment of the quality of the databases to be matched, or from a manual evaluation of an earlier matching of the same databases. Alternatively, these estimates can be calculated based on population estimates, like in the month of birth example given above. Herzog et al. [143] discuss in detail how the m_i and u_i parameters can be estimated using either data from prior data matching projects or by employing the unsupervised expectation–maximisation (EM) algorithm.

Extensions to the basic Fellegi and Sunter approach to probabilistic record linkage include allowing for approximate comparisons of attribute values that result in similarity values in the agreement patterns γ rather than only agreement and disagreement values. Porter and Winkler [215, 279, 286] showed that modifying the m- and u-probabilities in Eqs. 6.8 and 6.9 using the normalised similarities (between 0.0 and 1.0) calculated by approximate string comparison algorithms can lead to significant improvements in matching quality.

The second extension is concerned by taking the frequency of attribute values into account when calculating the m- and u-probabilities [108, 286]. The intuition behind this idea is that the more frequent an attribute value is in a database, the less discriminative this value is for classifying a record pair as a match or non-match. The example using the surname values 'Smith' and 'Dijkstra' given at the beginning of this section has already illustrated this issue. Match and non-match weights should be adjusted according to the frequency of occurrence of individual attribute values, with lower m-probabilities for more frequent attribute values [108]. Herzog et al. [143] provide a detailed discussion of how frequency-based matching parameters can be calculated.

Winkler developed a method that combines the traditional Fellegi and Sunter approach to probabilistic record linkage with Bayesian networks [281]. Bayesian networks [138] can model selected dependencies between attributes. In general they, however, require training data (in the case of data matching in the form of record pairs with known true match status). Using both labelled and unlabelled training data, a modification of the EM algorithm was used by Winkler to estimate parameter settings. Viewing probabilistic record linkage from a Bayesian perspective has also been discussed by Fortini et al. [112] and Herzog et al. [143].

6.4 Cost-Based Classification

In the traditional probabilistic record linkage approach the two thresholds t_l and t_u are set such that the overall number of misclassified candidate record pairs is minimised. Two types of errors can occur (as will be further discussed in Chap. 7). First, a pair of records that refers to the same real world entity (and therefore is a true match) is classified as a non-match. Second, a pair of records that refers to two different entities (and thus is a true non-match) is classified as a match. Traditionally, it is assumed both types of errors have the same costs.

In many data matching and deduplication applications, however, these two types of errors have different costs [129, 263]. For example, imagine a health application where patient data from several databases (that contain information, for example, about prescriptions, hospital admissions and doctor consultations) are matched. Assuming these databases were matched such that each patient in the matched database is assumed to have a serious illness based on their medical history. These patients are invited by the hospital for a series of special medical tests to confirm if they do have this illness or not. Testing a patient for this illness will incur a certain amount of money, possibly in the hundreds or even thousands of dollars. Therefore, each patient that has been matched falsely will mean an increase in costs for an additional test that might have been unnecessary. On the other hand, each patient that was not classified as a match but who potentially has this serious illness might die because they are not given the medical test that could confirm if they have the illness or not. The costs for such a missed true match can therefore be the loss of life of an individual.

Another, less dramatic, example can be found in marketing, where often databases are matched to generate mailing lists of potential customers who are interested in certain topics, based on their shopping history (like sporting, gardening, music or reading). The cost of sending an advertisement flyer about a certain topic to somebody who is not interested in this topic is very small, compared to not sending the flyer to somebody who will probably respond to the advertisement [263]. Missing such a customer can potentially result in a significant loss in profit.

Table 6.1 Costs associated with various matching decisions as proposed by Verykios et al. [263]

Cost	Classification	True match status
$c_{U,M}$	Non-Match	True match (M)
$c_{U,U}$	Non-Match	True non-match (U)
$c_{P,M}$	Potential Match	True match (M)
$c_{P,U}$	Potential Match	True non-match (U)
$c_{M,M}$	Match	True match (M)
$c_{M,U}$	Match	True non-match (U)

As shown in these two examples, clearly there can be different costs associated with false matches and false non-matches. A cost-optimal decision model based on a Bayesian approach has been developed by Verykios et al. [263]. In this approach, the decision rule for an agreement pattern γ in Eq. 6.7 is formulated in a Bayesian setting:

$$P(\gamma \in \Gamma | r \in M) \geq P(\gamma \in \Gamma | r \in U) \Rightarrow r \to \text{Match}, \qquad (6.12)$$
$$P(\gamma \in \Gamma | r \in M) < P(\gamma \in \Gamma | r \in U) \Rightarrow r \to \text{Non-match}.$$

As shown in Table 6.1, different costs can be assigned to each of the six decision outcomes in the traditional Fellegi and Sunter model, where record pairs are classified into matches, non-matches and potential matches. The objective of a cost optimal decision rule is then to minimise the overall cost c:

$$\begin{aligned}
c =\,& c_{U,M} \cdot P(r \in \text{Non-Match}, r \in M) + c_{U,U} \cdot P(r \in \text{Non-Match}, r \in U) + \\
& c_{P,M} \cdot P(r \in \text{Potential Match}, r \in M) + c_{P,U} \cdot P(r \in \text{Potential Match}, r \in U) + \\
& c_{M,M} \cdot P(r \in \text{Match}, r \in M) + c_{M,U} \cdot P(r \in \text{Match}, r \in U), \qquad (6.13)
\end{aligned}$$

where $P(x, y)$ is the joint probability that a record pair r has been classified into class x (with $x \in \{\text{Non-Match, Potential Match, Match}\}$) while the true match status of r is y (with $y \in \{M, U\}$). Bayes theorem can then be applied to replace these six probabilities with the probabilities of a certain match decision, given the true match status and the a priori probabilities of $P(M)$ and $P(U)$:

$$P(r = x, r = y) = P(r = x \mid r = y) \cdot P(r = y), \qquad (6.14)$$

with x and y being a value of the corresponding two sets given above. The probabilities $P(r = x \mid r = y)$ and $P(r = y)$ can both be estimated using training data that are available in the form of record pairs with known true match status [263]. An optimal decision rule, similar to the one given in Eq. 6.7, can then be developed, which for different values of the different costs provides an overall cost-optimal decision [263].

Cost-based classification is not just possible for the probabilistic record linkage approach as was presented in this section, but for other classification techniques for data matching as well. In rule-based classifiers (discussed next), rules can for example, be reordered such that the rules that classify candidate record pairs into matches are evaluated before rules that classify them into non-matches, while for many supervised machine learning classifiers different costs for different classes can be incorporated into the learning process.

6.5 Rule-Based Classification

A rule-based classification approach is different to the probabilistic approaches presented in the previous two sections. It employs rules that classify candidate record pairs into matches and non-matches (and maybe potential matches that are passed on for manual clerical review) [82, 141, 195]

A rule-based classifier can be applied on the similarity values of the comparison vectors generated in the comparison step. Rules are made of individual tests on certain similarity values that are combined with conjunctions (logical and), disjunctions (logical or) and negations (logical not). Figure 6.3 shows an example set of such rules.

The form of a rule is $P \Rightarrow C$, where P is a predicate that is applied on the similarity values (as available in a comparison vector) for a record pair (r_i, r_j), and C is the classification outcome of the pair (r_i, r_j). The predicate P is a boolean expression of the general form:

$$P = (term_{1,1} \vee term_{1,2} \vee \ldots) \wedge \ldots \wedge (term_{n,1} \vee term_{n,2} \vee \ldots). \qquad (6.15)$$

P is written in conjunctive normal form as a conjunction of disjunctions of terms [195]. Each term is a test applied on the similarity value of a single element in a comparison vector of the record pair (r_i, r_j). For example, a term can be a test such as $s(\text{GivenName})[r_i, r_j] \geq 0.7)$, i.e. if the similarity value for the record pair r_i and r_j for the given name attribute is equal to or greater than 0.7. In Fig. 6.3, each disjunction only contains one term.

The classification outcome C of a rule assigns a candidate record pair into the class given in C when a rule is triggered (i.e. when the predicate P is true). A rule system can either consist of rules that classify record pairs into matches only, or of rules that classify pairs into matches, non-matches and even potential matches. In the first case, all record pairs that are not covered by any rule will implicitly be classified as non-matches. For the second case, a rule set needs to cover all possible values in the similarity values of the comparison vectors, as there is no default class, or a default class needs to be set explicitly (for example in the form of a rule with an empty predicate P and where C classifies a record pair as a non-match).

If a rule set only contains rules that classify candidate record pairs as matches, then the ordering of these rules is irrelevant. On the other hand, if a rule set consists of rules that classify record pairs into more than one class, then the ordering of rules is crucial. For a given record pair, the first rule where the predicate P becomes true (the first rule that is triggered or 'fired') is the rule that classifies the pair.

$$(s(\text{GivenName})[r_i, r_j] \geq 0.9) \;\wedge\; (s(\text{Surname})[r_i, r_j] = 1.0)$$
$$\wedge\; (s(\text{BMonth})[r_i, r_j] = 1.0) \;\wedge\; (s(\text{BYear})[r_i, r_j] = 1.0) \;\Rightarrow\; [r_i, r_j] \rightarrow \text{Match}$$

$$(s(\text{GivenName})[r_i, r_j] \geq 0.7) \;\wedge\; (s(\text{Surname})[r_i, r_j] \geq 0.8)$$
$$\wedge\; (s(\text{BDay})[r_i, r_j] = 1.0) \;\wedge\; s(\text{BMonth})[r_i, r_j] = 1.0)$$
$$\wedge\; (s(\text{BYear})[r_i, r_j] = 1.0) \;\Rightarrow\; [r_i, r_j] \rightarrow \text{Match}$$

$$(s(\text{GivenName})[r_i, r_j] \geq 0.7) \;\wedge\; (s(\text{Surname})[r_i, r_j] \geq 0.8)$$
$$\wedge\; (s(\text{StrName})[r_i, r_j] \geq 0.8) \;\wedge\; (s(\text{Suburb})[r_i, r_j] \geq 0.8) \;\Rightarrow\; [r_i, r_j] \rightarrow \text{Match}$$

$$(s(\text{GivenName})[r_i, r_j] \geq 0.7) \;\wedge\; (s(\text{Surname})[r_i, r_j] \geq 0.8)$$
$$\wedge\; (s(\text{BDay})[r_i, r_j] \leq 0.5) \;\wedge\; (s(\text{BMonth})[r_i, r_j] \leq 0.5)$$
$$\wedge\; (s(\text{BYear})[r_i, r_j] \leq 0.5) \;\Rightarrow\; [r_i, r_j] \rightarrow \text{Non-Match}$$

$$(s(\text{GivenName})[r_i, r_j] \geq 0.7) \;\wedge\; (s(\text{Surname})[r_i, r_j] \geq 0.8)$$
$$\wedge\; (s(\text{StrName})[r_i, r_j] \leq 0.6) \;\wedge\; (s(\text{Suburb})[r_i, r_j] \leq 0.6) \;\Rightarrow\; [r_i, r_j] \rightarrow \text{Non-Match}$$

Fig. 6.3 An example set of classification rules that could be applied on the comparison vectors from Fig. 2.6 shown on page 31. Conjunctions (logical and) are shown as $\wedge$ and disjunctions (logical or) as $\vee$. '$s(\cdot)$' refers to a similarity value taken from the comparison vector for a given record pair. The first three rules classify a pair of records r_i and r_j as a match if their name values are similar, and either their dates of birth or their addresses are similar as well. On the other hand, the last two rules classify a pair as a non-match if their name is similar but they have either a different date of birth or a different address

Ideally, each rule in a set of rules should be of high accuracy and high coverage [135]. A high accuracy means that a rule that classifies record pairs into a certain class should mostly cover pairs that do belong to this class but not pairs that belong into another class. In order to be able to assess the accuracy of rules, candidate record pairs and their true match status (match or non-match) must be available. Without the true match status it is not possible to assess the accuracy of rules. A high coverage means that the predicate P of a rule covers a large portion of all candidate record pairs. A rule which has a coverage of 10 % is triggered (i.e. its predicate P is true) for 10 % of all candidate record pairs. A rule with an empty predicate P (i.e. no test on any similarity value) would have a coverage of 100 %. The more specific a rule is (i.e. the more conditions are tested in the predicate P) the lower the coverage of a rule generally becomes. More specific rules are usually more accurate, while less specific rules often have lower accuracy because they cover a larger number of candidate record pairs that are in both the match and non-match classes.

The quality of a rule set can be measured by its overall accuracy and its coverage [135]. Additionally, a smaller rule set is generally preferable over a larger rule set, because a smaller number of rules is easier to maintain. Because rules are depending upon the characteristics and the quality of the data that are matched or deduplicated, either a new set of rules needs to be developed for each new database, or an existing set of rules needs to be adjusted when data with different characteristics are matched or deduplicated.

This raises the question of how a set of rules can be generated to achieve a high classification accuracy of the compared candidate record pairs. The two basic approaches are to either develop a rule set manually or to learn a set of rules from training data.

- The traditional approach to generating rules is to manually develop them based on domain knowledge of the databases to be matched or deduplicated. Developing such rules is usually done hand in hand with selecting appropriate indexing approaches and comparison functions, because both of these will affect the candidate record pairs that are generated and the similarity values in their corresponding comparison vectors.

  Manually generating rules is a labour-intensive process that is generally iterated over many variations of potential rules. These rules need to be tested and manually evaluated using some form of training data that contain the true match status of candidate record pairs. If such training data are not available (as is the case in many real world data matching situations, as will be discussed further in Chap. 7), then the evaluation of each rule requires manual inspection of all candidate record pairs that are covered by a rule, and for each covered pair it needs to be manually decided if the classification is correct or not. This is a tedious and labour-intensive process.

- An alternative approach to generating a set of rules is to learn them from training data that consist of candidate record pairs and their true match status. Similar to the learning of blocking keys discussed in Sect. 4.12, the learning of rules can be accomplished by employing a sequential covering algorithm [135], where a set of rules that cover one class (usually the candidate record pairs that correspond to matches) is learned first, followed by rules that cover the other class (the non-matches).

  One rule is learned after another, by starting with an empty predicate P for a rule and evaluating its accuracy and coverage. Candidate rules are then generated by adding a term to P based on the similarity values in the different elements of comparison vectors. Such candidate rules could for example be (similar to the example given in Fig. 6.3):

$$(s(\text{GivenName})[r_i, r_j] = 1.0) \Rightarrow [r_i, r_j] \rightarrow \text{Match}$$
$$(s(\text{Surname})[r_i, r_j] = 1.0) \Rightarrow [r_i, r_j] \rightarrow \text{Match}$$
$$(s(\text{StrName})[r_i, r_j] = 1.0) \Rightarrow [r_i, r_j] \rightarrow \text{Match}$$
$$(s(\text{Suburb})[r_i, r_j] = 1.0) \Rightarrow [r_i, r_j] \rightarrow \text{Match}$$

The best candidate rule (according to some criteria that takes accuracy and coverage into account [135]) is selected, and this becomes the new base rule which will be expanded with new candidate terms in the next step [135]. Assuming for example, that the first of the four above rules was the best candidate, the next set of expanded candidate rules could consist of:

$$(s(\text{GivenName})[r_i, r_j] = 1.0) \land (s(\text{Surname})[r_i, r_j] \geq 0.8) \Rightarrow [r_i, r_j] \rightarrow \text{Match}$$
$$(s(\text{GivenName})[r_i, r_j] = 1.0) \land (s(\text{StrName})[r_i, r_j] \geq 0.8) \Rightarrow [r_i, r_j] \rightarrow \text{Match}$$
$$(s(\text{GivenName})[r_i, r_j] = 1.0) \land (s(\text{Suburb})[r_i, r_j] \geq 0.8) \Rightarrow [r_i, r_j] \rightarrow \text{Match}$$

This process of testing candidate rules and expanding the best candidate with another term is repeated until a stopping criteria is fulfilled. All candidate record pairs that are covered by the latest generated rule are then removed from the training set, and if candidate record pairs are left in the training set then a new rule is learned.

Two data matching research prototypes were developed in the late 1990s that were employing rule-based classification approaches. A system based on an extension of SQL that allows rule-based matching operators to be defined was proposed by Galhardas et al. [117]. These matching operators included similarity predicates, the setting of thresholds, as well as normal SQL statements. Complex matching statements were therefore written using SQL statements. A related approach was the WHIRL system developed by Cohen [81], which combined similarity calculations based on cosine similarity (described in Sect. 5.8) with conjunctive rules applied on record attributes.

More recently, Schewe and Wang [236] proposed a reasoning approach to acquire knowledge about entities stored in different databases by identifying objects through knowledge patterns. Such patterns can capture details such as abbreviations and variations in title, name and address values. An advantage of knowledge patterns is that they can capture knowledge at different levels of abstractions (i.e. not just at the level of individual entities but also at the level of attributes), and by using the contexts of where patterns occur (i.e. taking the relations between different patterns into account). Knowledge patterns allow a user to identify, for example, the types of name and address variations that commonly occur in two databases that are to be matched, which in turn can facilitate the development of rule-based classifiers that determine if two records correspond to the same entity or not depending upon the variations in their attribute values.

6.6 Supervised Classification Methods

When the compared candidate record pairs are only classified into matches and non-matches (but not potential matches), then this classification is known as a *binary classification problem*. Further, if training data in the form of record pairs with their true match status (match or non-match) are available, then a supervised classification approach can be employed to train a classification model using these training data. The trained model is then used to classify record pairs with an unknown match status into matches and non-matches. Many binary classification techniques have been developed by the AI, machine learning and data mining communities over the past few decades [135, 189], and several of these techniques have been applied in the area of data matching and deduplication. This section provides an overview of this work and highlights important issues that need to be considered when a supervised classification technique is used to classify record pairs.

Most classification techniques (including the probabilistic record linkage, cost-based and rule-based approaches discussed earlier in this chapter), classify each compared record pair individually and independently from all other record pairs (Sect. 6.10 below covers techniques that are aimed at classifying all compared record pairs in a collective approach). From the classification point of view, each compared record pair is represented by its comparison vector that contains the individual similarity values that were calculated in the comparison step (as was discussed in Sect. 5.16). These comparison vectors correspond to the *feature vectors* (the notation used in machine learning or data mining) that are employed to train a classification model, and to classify record pairs with unknown match status. Figure 6.4 shows such a set of comparison vectors and their true match status. A supervised classification approach consists of three steps [135].

1. A supervised classification technique is selected and a classification model is built by training the classifier using available training data which include the known true match status of candidate record pairs. Overviews of supervised classification techniques are provided in text books on machine learning and data mining [135, 189]. Most classification techniques require a user to tune a variety of parameters to achieve high classification accuracy. Selecting appropriate parameter values can be conducted either via a guided search through the parameter space or via manual tuning.

2. The accuracy of the built classification model is evaluated using a set of testing data that must be in the same format and structure as the training data (i.e. these data must be comparison vectors that were generated using the same comparison functions as the comparison vectors in the training data). These testing data must also contain the known true match status of record pairs, so that the match or non-match decision of the trained classifier can be compared with the true match status (this topic is covered in more detail in Sect. 7.2).

   It is important that the testing data are different from the training data, because otherwise *over-fitting* can occur [135]. Over-fitting refers to the issue that the accuracy of a classification model as measured on the training data is very high, because the model will learn the intrinsic characteristics of the training data. Testing a model on data sets that are different from the training data set is more meaningful and more realistic, because in practice the data upon which a classifier is applied on will be different from the data the classifier was trained on. The accuracy reported using a testing data set is therefore closer to the accuracy that can be expected when the classification model is applied on new, unseen data where the match status of candidate record pairs is unknown.

   If the accuracy reported on the testing data is not good enough according to some criteria set for a certain data matching exercise, then one needs to go back to step 1 and either change some of the parameter settings used when the classifier was trained, or alternatively employ a different classification technique altogether.

RecPairID	GivenName	Surname	StrNum	StrName	Suburb	BDay	BMonth	BYear	Class
(a1,b1)	0.6	0.8	0.0	1.0	0.6	0.5	0.5	1.0	M
(a1,b2)	0.0	0.15	0.0	0.5	0.0	0.5	0.0	0.75	U
(a2,b1)	0.2	0.0	0.0	0.1	0.15	0.0	0.0	0.75	U
(a2,b2)	0.0	0.25	1.0	0.4	0.6	1.0	1.0	0.75	M

Fig. 6.4 Example comparison vectors based on records shown in Fig. 6.1 and their true match status (the column 'Class', where M corresponds to matches and U corresponds to non-matches), which can be used to train a supervised classifier

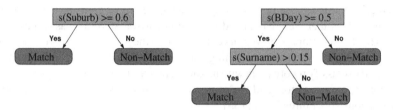

Fig. 6.5 Two example decision trees resulting from the four training comparison vectors from Fig. 6.4. The tests are conducted on the similarity values (indicated by 's($\cdot$)' in the tree nodes) for certain attributes. The leaf nodes correspond to the two classes into which record pairs are classified. Clearly, the tree on the left side is better as it is not only smaller (less tests) and thus faster when new candidate record pairs with unknown match status are to be classified, but also more robust. The test 's(Surname) > 0.15', which is a very low threshold, is unlikely to lead to accurate matching results. Rather, the tree on the right-hand side is over-fitting the training data shown in Fig. 6.4

It is important to note that the selection of the best classification technique is dependent on the data that are to be classified [80]. For different types of data different techniques might perform best.

3. Once a satisfactory accuracy has been achieved with a trained classification model, in the third step the model is applied to classify new unseen data, i.e. comparison vectors that correspond to candidate record pairs where the match status is not known.

In the following, two popular supervised classification techniques that have been employed in the area of data matching and deduplication are described in more detail.

- *Decision tree induction*: Decision tree induction is one of the most popular supervised classification techniques used in data mining and machine learning [135]. Decision trees, as the example shown in Fig. 6.5 illustrates, are favoured by many researchers and practitioners over other techniques because they can be visualised easily and are thus understandable even by people who are not data mining or machine learning experts. Additionally, decision trees can be directly converted into a set of rules. The two trees from Fig. 6.5 for example, can be converted into the following two sets of rules:

$$(s(\text{Suburb})[r_i, r_j] \geq 0.6) \Rightarrow [r_i, r_j] \to \text{Match}$$
$$(s(\text{Suburb})[r_i, r_j] < 0.6) \Rightarrow [r_i, r_j] \to \text{Non-Match}$$

$$(s(\text{BDay})[r_i, r_j] \geq 0.5) \wedge (s(\text{Surname})[r_i, r_j] > 0.15) \Rightarrow [r_i, r_j] \to \text{Match}$$
$$(s(\text{BDay})[r_i, r_j] \geq 0.5) \wedge (s(\text{Surname})[r_i, r_j] \leq 0.15) \Rightarrow [r_i, r_j] \to \text{Non-Match}$$
$$(s(\text{BDay})[r_i, r_j] < 0.5) \Rightarrow [r_i, r_j] \to \text{Non-Match}$$

Like with the rule-based classification approach described in the previous section, each internal node of a decision tree corresponds to a test on a similarity value in a comparison vector for a certain attribute, as illustrated in Fig. 6.5. Each internal node therefore corresponds to a test in the predicate of a rule, where the leaf nodes in a tree correspond to the possible classification outcome of a rule. In the case of data matching, the two possible outcomes are the match and non-match classes.

In the learning phase, a tree is built recursively, starting with an empty tree. At each step in the tree generation process, an attribute that results in the purest split of the training data set is selected (such that matches are moved into one branch of the tree and non-matches into the other branch). Different decision tree algorithms and splitting criteria have been developed. The interested reader is referred to text books in machine learning or data mining, such as the ones by Mitchell [189] or Han and Kamber [135].

An early work that used a decision tree classifier for data matching was presented by Cochinwala et al. [80]. Their work aimed at matching two databases with customer records. They manually generated training data in the form of sampled pairs of records that were then used to train a Classification and Regression Tree (CART) classifier [42]. Once a tree was generated, they applied tree pruning in order to reduce the complexity of the rules that were extracted from the tree, and to make these rules more robust. The reduced tree not only generated less complex rules (i.e. rules made of a smaller number of tests), it also lead to rules that were more robust when applied to matching the full customer record databases [80].

Elfeky et al. implemented the ID3 decision tree algorithm into their TAILOR data matching tool box [102]. They provided two approaches to generate training data. In the first approach, selected candidate record pairs are manually classified as matches and non-matches by a domain expert, and the comparison vectors of these record pairs are then used to train a decision tree. The second approach aims to overcome the manual step by first applying a clustering technique to group all candidate record pairs into three clusters based on their comparison vectors. The first cluster corresponds to the class of matches, the second cluster to the class of non-matches, and the third cluster to the class of potential matches. The comparison vectors in the match and non-match clusters are then used to train the decision tree classifier. In their experimental study, the authors found that both the decision tree based on manual training data generation and the one based on the cluster pre-processing (called the 'hybrid classification approach') achieved better matching quality than the threshold-based probabilistic record linkage classifier described in Sect. 6.3.

- *Support vector machine (SVM)*: This relatively recent classification technique, developed in the 1990s [259], is based on the idea of mapping the training data set, which consists of comparison vectors and their class labels (match or non-match), into a multi-dimensional vector space in such a way that the training records from the two classes are separated and the gap between the two classes is made as wide as possible.

  This idea is illustrated in Fig. 6.6. A decision boundary corresponds to a hyperplane in the high-dimensional space (a line in two dimensions or a plane in three dimensions), and the optimal decision boundary is the one which has the widest margins to training records in both classes. The mapping from the original input space (i.e. the comparison vectors containing similarity values) into a high-dimensional space is conducted using a *kernel function*, which allows the efficient calculation of the dot product required in the training process of a SVM. This training process corresponds to solving a quadratic optimisation problem [259], for which efficient techniques are available.

  Bilenko et al. [35] employed a SVM classifier to learn the costs for edit operations (such as character inserts, deletes or substitutions) within the Levenshtein edit distance approximate string comparison function (which was presented in Sect. 5.3). Learning these costs allows a better separation of the string pairs that correspond to matches from those that correspond to non-matches. The training data required for this approach consist of pairs of strings and their match status.

  Christen [59, 60] developed an automatic classification approach for data matching based on a SVM, which is similar to the clustering-based hybrid approach developed by Elfeky et al. [102] described above. In a first step, training examples that clearly correspond to matches and non-matches are selected from the set of all comparison vectors. Clear match examples are comparison vectors where all similarity values are equal to or very close to the exact similarity of 1, while clear non-match examples are comparison vectors where all similarity values are equal to or close to 0. Based on this initial training set, a first SVM is trained. All comparison vectors that are not in one of the two training sets are classified using this initial SVM. In the second step, the comparison vectors that were classified to be furthest away from the SVM decision boundary are added into one of the two training sets (depending upon if they are located on the side of matches or on the side of non-matches), and a second SVM is trained on these enlarged training sets. This process of adding more comparison vectors into the training sets followed by training a new SVM is repeated until a stopping criteria is fulfilled. In an experimental study, this automatic classification approach outperformed a basic clustering approach as well as the hybrid approach by Elfeky et al. [102] in experiments on several data sets.

Employing supervised classification techniques for data matching has several challenges. First, classifying candidate record pairs is often an imbalanced problem, in that there are many more record pairs that correspond to true non-matches com-

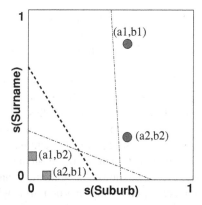

Fig. 6.6 A simplified illustration of a 2D vector space (made of the similarities of the attributes 'Surname' and 'Suburb'), containing the similarity values of the four comparison vectors from Fig. 6.4, and three decision boundaries (*dotted lines*) that correspond to three trained support vector machine (SVM) classifiers. The *thick dotted line* is the SVM which has the widest margins to both the class of matches (*circles*) and non-matches (*squares*)

pared to the number of record pairs that correspond to true matches. This holds even after some form of indexing has been applied. As a result, a classification technique, as well as the measure(s) used to evaluate how good a trained classification model is, must be able to handle imbalanced classes. The way training data are generated can help to overcome this problem, for example by sampling the same number of training examples from both the match and non-match classes. This issue will be discussed further in Sect. 7.1.

The second issue is the difficulty to generate, obtain, select or sample training data that are representative of the actual data that are to be matched. Good training data will more likely result in a robust and accurate classification model. Acquiring or manually generating training data can be quite costly and time-consuming. As a result, training data sets are often small compared to the databases that are to be matched or deduplicated. Training data should, however, represent the detailed characteristics of the full database(s) as much as possible. An alternative to train a supervised classifier using a large training data set is to create training data interactively using an active learning approach, as will be discussed next.

6.7 Active Learning Approaches

A major drawback of supervised classification techniques is their need for training data sets, made of comparison vectors that correspond to matches and non-matches, that represent the characteristics of the full database(s) to be matched or deduplicated. An alternative to generating or obtaining such comprehensive training data sets is to use a classification approach that only requires a small amount of training data in order to achieve high classification accuracy. Based on an initial small training data set, a classification model is built interactively by asking an experienced user for further training examples that help to improve the classification model. Such interactive approaches are known as *active learning* [11, 231, 252].

An active learning classifier starts by building a first classification model using a small set of *seed* training examples. These can, for example, be comparison vectors that correspond to clear matches and clear non-matches. This initial classification model will likely have a low classification accuracy. Specifically, it will have difficulties to classify comparison vectors with certain characteristics, such as comparison vectors that do not correspond to clear matches or non-matches. If a SVM classifier is used, for example, then the comparison vectors that are located closest to the decision boundary (see Fig. 6.6) correspond to matches or non-matches with almost equal likelihood. A manual classification of the candidate record pairs that correspond to these comparison vectors can be highly beneficial to improve the accuracy of the classification model.

An active learning classifier works iteratively by (1) training a classification model, (2) classifying all comparison vectors not in the training set as matches or non-matches, (3) asking a user to provide manual classification of the candidate record pairs that were most difficult to classify, (4) adding these manually classified comparison vectors to the corresponding training data set (of either matches or non-matches) and (5) training the next, improved, classification model. This process is repeated until a certain stopping criteria is met. This stopping criteria either terminates this process after a maximum number of iterations, or more commonly when the last trained classifier achieves a certain matching quality on the testing data set.

The following three classification approaches using active learning have been proposed in the area of data matching and deduplication.

- Sarawagi et al. [231] presented the ALIAS system, which is an interactive deduplication system that (similar to the traditional probabilistic record linkage classification approach discussed in Sect. 6.3) works with the three classes of matches, non-matches and potential matches. Rather than building only one classification model, a set of several models is trained on the training data set, each of them with a randomised choice of parameter setting. Three decision trees were used in ALIAS. For those comparison vectors where different decisions were made by the three trained decision trees (for example, two classify a comparison vector as a match and one as a non-match), a manual decision is required by the user. According to this manual classification, comparison vectors are added into either the training set of matches or the set of non-matches, and the next set of classifiers is trained on this enhanced data set.

- A similar approach was presented by Tejada et al. [252] aimed at integrating data objects from different Web sources. Their system, called Active Atlas,[1] learns *mapping rules* using an active learning approach. These mapping rules include tests for string equality, string prefix or string suffix equality, or if two strings

---

[1] It is interesting to note that both the ALIAS and Active Atlas systems were presented in the same year (2002) and at the same conference.

contain the same abbreviations or acronyms (like 'IBM' vs. 'International Business Machines'). Similar to the ALIAS system, a committee of three decision tree classifiers was used to learn the rules that best distinguish matches from non-matches, with a manual classification required for pairs of strings where the three decision trees returned different classification outcomes.

- More recently, Arasu et al. [11] presented a novel approach to active learning specifically designed for data matching. Their technique integrates indexing with active learning. Either a decision tree or SVM classification model can be employed, and a user can specify the minimum precision (to be discussed in Sect. 7.2) the final classification model must achieve. The active learning process then aims to achieve a high recall for the classification model while reducing the number of examples to be classified manually as much as possible. Experiments on two large databases showed that this proposed new technique outperformed both ALIAS [231] and Active Atlas [252].

6.8 Managing Transitive Closure

The result of the classification of individual candidate record pairs into matches and non-matches is often not the final outcome of a data matching or deduplication exercise. If candidate record pairs are classified individually, each record can be part of a match with several other records, as illustrated in Fig. 6.7. In certain situations, however, a one-to-one match restriction has to be applied, as will be further discussed in Sect. 6.11. If multiple matches are allowed, then the issue of *transitive closure* needs to be addressed.

Transitive closure refers to the situation where two record pairs, (r_i, r_j) and (r_i, r_k), have been classified as matches but the pair (r_j, r_k) has been classified as a non-match. This contradicts the intuition that if record r_i is considered to be a match with record r_j (i.e. referring to the same entity) and record r_j is considered to be a match with r_k, then record r_i must also be considered a match with r_k. Applying the transitive closure refers to changing the match status of record pairs such that no contradictions of the match status within groups of records occurs [195].

The transitivity of matches can also lead to problems in that 'chains' of records, where individual pairs are classified as matches, are formed. The records at the two ends of a chain can, however, be quite different from each other, and they would not be considered to correspond to a match. For example, consider the four records 'a1' to 'a4' in Fig. 6.7, where the three individual pairs (a1,a2), (a2,a3) and (a3,a4) have been classified as matches, but the summed similarities between other pairs is below the match classification threshold $t = 5$. Pair (a1,a4), for example, only has a summed similarity of $SimSum(a1, a4) = 1.15$, and it is unlikely that these two records refer to the same individual. The clustering approaches discussed in the following section aim to overcome this problem of chains of matching records.

In real-world databases, the problem of record chains being generated by a pairwise classification technique seems to occur only rarely because the space of all

RecID	GivenName	Surname	StrNum	StrName	Suburb	BDay	BMonth	BYear
a1	john	smith	18	miller st	dickson	12	11	1970
a2	jonny	smith	73	miller st	dixon	11	10	1970
a3	joan	smith	73	dawson cr	lyneham	11	12	1979
a4	max	miller	73	dawson cr	lyneham	11	2	1969
a5	sal	bass	67	milles rd	ainslie	28	5	1981
a6	sally	bass	64	miles rd	ainsile	23	5	1981

Candidate pair	SimSum	Classification
(a1, a2)	5.20	Match
(a1, a3)	3.30	Non-match
(a1, a4)	1.15	Non-match
(a2, a3)	5.05	Match
(a2, a4)	2.70	Non-match
(a3, a4)	5.25	Match
(a5, a6)	6.20	Match

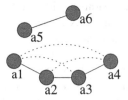

Fig. 6.7 An example of the transitive closure problem for a set of six records (top row table, 'a1' to 'a6'). It is assumed the summed nonzero similarities SimSum for the candidate record pairs in the lower left table have been calculated, and a simple classification threshold-based classifier with threshold $t = 5.0$ has been applied to classify each record pair individually. The result of this classification are two groups of records (possibly corresponding to two entities). The transitive closure would require that the record pairs (a1,a3), (a1,a4) and (a2,a4) are also considered to be matching (illustrated with dotted lines), even though their summed similarities are below the classification threshold

possible values in the different record attributes is very large. The likelihood that records that are not matching have a high similarity with each other is therefore very small [140, 190, 191]. Clustering and collective classification algorithms, which will be discussed in the following two sections, deal with the problem of transitivity by default, by classifying groups of records as matches rather than individual pairs of records only.

6.9 Clustering-Based Approaches

So far all techniques presented have viewed the problem of deciding which candidate record pairs correspond to matches and which to non-matches as a traditional classification problem. An alternative approach is to view this classification as a clustering (grouping) approach, where each cluster consists of records that refer to one entity. Clustering is the process of grouping data objects that are similar to each other according to some criteria into the same cluster [135]. The aim of clustering is to generate clusters that have high intra-cluster similarity and low inter-cluster similarity. This means all data objects within a cluster should be similar with each other, while data objects in different clusters should be dissimilar to each other.

Clustering is generally conducted in an unsupervised fashion, and therefore no training data in the form of record pairs with known true match status are required. Many different clustering techniques have been developed by the statistics, machine learning and data mining communities [135]. Different clustering techniques employ different heuristics to guide the clustering process [135]. They either partition data objects into a fixed number of clusters or into a hierarchy of clusters, or they generate graphs that correspond to clusters (to be discussed in more detail below), or they generate clusters that correspond to dense areas where many data objects are located close to each other. In data matching, the data objects to be clustered correspond to the records that represent entities.

A clustering-based approach is clearly suited for deduplication, where all records to be matched are stored in one database. For the matching of two or more databases, however, all records first need to be inserted into one common set. This can be accomplished by assigning each record a unique identifier which not only identifies the record but also the source database from where the record has originated.

Because each entity in a database will ideally be represented by one cluster, the number of clusters is not only unknown at the beginning of a clustering process, it will also be very large when the databases that are deduplicated or matched contain many entities. Partitioning based clustering algorithms [135], which require the number of clusters to be specified at the beginning, are therefore not applicable for clustering records in data matching or deduplication applications.

The clusters generated in data matching are generally very small, containing only a few records. Some, potentially many, clusters will only consist of a single record, if there is only one record in the database(s) that corresponds to this entity.

Different clustering approaches for data matching and deduplication have been investigated. In the following, five different approaches are discussed in more detail.

- In an early clustering approach, Monge [190] proposed an adaptive deduplication system where records are clustered according to some similarity measure, and a priority queue is kept in memory consisting of the most recently formed clusters. Each cluster corresponds to an entity, and is made of one or several records that represent this entity. To save memory, however, not necessarily all records that refer to an entity are kept in memory for a given cluster. Initially all records to be deduplicated or matched are sorted according to a sorting key (as was discussed in Sects. 4.2 and 4.5). One record in the sorted database is then processed after another. Each record is compared with the records stored in the priority queue. If a match is found the current record is attached to the matching cluster, and this cluster is put at the top of the priority queue. If no match is found then a new cluster is formed made of the current record only. To make sure that only a certain amount of memory is used, the oldest cluster is removed from the end of the priority queue if a new cluster is generated and the queue exceeds a maximum length limit.

  The experimental results of this combined sorted-neighbourhood and clustering technique presented by Monge showed that the approach can achieve matching accuracies similar to the basic sorted-neighbourhood approach [140, 141]. It can, however, reduce the number of record pair comparisons that are conducted by upto

75 %, because each record is only compared to a small number of representative records in a cluster.

- Clustering can also be applied as a post-processing step after the pair-wise classification of record pairs has been conducted, and a graph of all matching records, as for example shown in Fig. 6.7, has been generated. The aim of clustering using such a graph is to decide for each sub-graph (consisting of connected records) which record subsets correspond to the actual entities that are to be matched or deduplicated [140, 195]. For example, the sub-graph made of the four records 'a1' to 'a4' from Fig. 6.7 is unlikely to refer to one but rather to three entities (only 'a1' and 'a2' seem to be duplicate records of the same individual).

  One approach to reducing the size of sub-graphs (and thus the number of records that can correspond to the same entity) is to iteratively remove edges between two nodes (corresponding to a record pair classified as a match) starting from the edge that has the lowest similarity in a sub-graph. For the sub-graph made of records 'a1' to 'a4' in Fig. 6.7, the edge from 'a2' to 'a3' has the lowest similarity (5.05), therefore this edge would be removed first, leaving two new smaller sub-graphs. This process can be repeated until either each sub-graph only contains edges with a certain minimum intra-cluster similarity value t_c, until the transitive closure property has been fulfilled, or alternatively until each sub-graph contains no more than a maximum number n_c of records [195]. Which of these stopping criteria is best suited depends upon the requirements of the data matching or deduplication application.

  Continuing on with the example from Fig. 6.7, if the minimum intra-cluster threshold is set to $t_c = 5.25$, then the link between records 'a2' and 'a3' is removed first. The link between records 'a1' and 'a2' will also be removed, resulting in three separate entities (which possibly corresponds to a missed true match), while the link between records 'a3' and 'a4' is kept as a match (possibly a wrong match). On the other hand, if the maximum size of a sub-graph is set to $n_c = 2$, then the record pair 'a1' and 'a2' is considered as one entity and the pair 'a3' and 'a4' as another entity, and only the link between 'a2' and 'a3' is removed.

- Another approach to clustering a graph of matching record pairs is to find centres within each sub-graph and to then assign nodes (records) to their closest centre, i.e. the centre record they are most similar to. This approach, named CENTER [137], first sorts the edges of a sub-graph in descending order of their similarities. The first time a record r_i, appears in an edge of the sub-graph it is assigned as the centre of a cluster. All records r_j that appear in edges (r_i, r_j) later on in the sorted list are then assigned to this cluster, but not to any other clusters [195].

  When this clustering technique is applied on the sub-graph of records 'a1' to 'a4' from Fig. 6.7, the sorted list of edges is: (a3,a4), (a1,a2) and (a2,a3), with similarities 5.25, 5.20 and 5.05, respectively. If node 'a3' is marked as the centre then 'a4' is obviously considered to be part of this cluster. In the next e.g., (a1,a2), neither 'a1' nor 'a2' have been marked as centres or as being part of a cluster, and so 'a1' is marked as a new centre. In the third e.g., (a2,a3), both nodes have already been assigned to clusters and so this edge is not considered. As a result, the clustering of this sub-graph leads to two sub-graphs that correspond to two entities, one consisting of records 'a1' and 'a2' and the other of records 'a3' and 'a4'. The selection of which node in a pair becomes the centre of a new cluster obviously affects the final clustering outcome. One approach to overcoming this problem, called MERGE-CENTER [137], is to merge two clusters if their centres are very similar to each other.

- The two previously described approaches to clustering are based on a graph of matching candidate record pairs which was built using a pair-wise comparison and classification technique. A drawback of these approaches is that the minimum similarity threshold that has been used to classify record pairs into matches and non-matches is determining the structure of the cluster graph. This threshold is a global parameter applied to all compared record pairs. An alternative approach is to cluster the set of records based on all similarity values calculated between pairs of records (not just the ones classified as matches), and to guide the clustering based on records that are similar to each other relative to the number of records that are located in the neighbourhood around them [195]. This is an approach similar to density based clustering [135].

  Chaudhuri et al. [52] proposed such an approach based on the concepts of compact sets and sparse neighbourhoods. A compact set, CS, is a group of records that are all more similar with each other (i.e. have small distances $dist(\cdot)$ with each other) than they are similar to any other records. Specifically, for all pairs of records $r_i, r_j \in CS : dist(r_i, r_j) < dist(r_i, r_k) \; \forall \; r_k \notin CS$. The neighbourhood set of a record r_i is defined as $N(r_i) = p \cdot nn(r_i)$, where $nn(r_i)$ is the distance of record r_i to its closest neighbour and p determines the size of the radius around r_i that is considered. The neighbourhood of r_i is defined to be sparse if the number of records in the set $N(r_i)$ is below a certain constant threshold [52]. The advantage of this clustering approach is that clusters of records are generated depending upon the number and density of their neighbouring records, rather than based on a global threshold.

- A different clustering approach was proposed by Verykios et al. [261] and Elfeky et al. [102]. Rather than clustering the actual records based on the similarities calculated between them, clustering was applied on the comparison vectors that are generated in the comparison step. Specifically, comparison vectors were inserted into three clusters, one each corresponding to matches, non-matches and potential matches, similar to the traditional probabilistic record linkage approach presented in Sect. 6.3. Identifying the clusters that correspond to matches and non-matches is easy because they will either have a centroid vector that is close to an exact match (with comparison vector $[1.0, \ldots, 1.0]$) or a centroid vector that is close to a total non-match (with comparison vector $[0.0, \ldots, 0.0]$), respectively. In the second step of this approach, the comparison vectors in the match and non-match clusters were used as training data for a decision tree classifier, as was previously discussed in Sect. 6.6.

6.10 Collective Classification

Pair-wise classification techniques make a match or non-match decision independently for each compared candidate record pair, and clustering techniques further refine the classification of groups of records that likely correspond to the same entity. With both these approaches, decisions about the match status of a pair or a group of records are made independently from all other records or groups in the database(s) that are matched or deduplicated. These techniques therefore make local decisions without taking the characteristics of all records in the full database(s) into account.

New techniques have been proposed in the past few years that aim to make a decision about which records are matching in an overall collective fashion over all pairs or groups of records in the database(s) that are matched. These techniques are known as 'collective entity resolution' techniques, and they employ either iterative or hierarchical clustering [31, 181], or graph-based approaches [93, 155, 195]. All of these collective classification approaches have been developed for, and evaluated on, databases that contain different types of entities, where certain relationships between entities are known. These relationships can be represented in a *relationship graph*, as illustrated in Fig. 6.8. The most popular type of such data are bibliographic databases where the entity types include *authors*, *institutions* (or affiliations), *venues* (journals, conferences and workshops), and the actual *papers* (or articles) [31, 155].

The basic idea of collective classification approaches is to calculate the similarities of all connections (links) in the relationship graph that are ambiguous (such as the dotted links in Fig. 6.8) using information from the known relationships (the 'hard' connections between different entities). Because different types of entities are available, the known relationships between one type of entities can help to disambiguate (i.e. decide the match status) of other types of entities.

The first step in collective classification techniques is to generate the relationship graph, which can consist of relations between different types of entities. These relations can either be 'hard' connections (where a relationship is known without doubt from the data), or connections that have a probability or weight attached to them if it is not clear if a relationship exists or not. These probabilities or weights can, for example, be based on similarities calculated when pairs of records are compared, as was discussed in Chap. 5. The collective classification task is then to decide if these possible relationships correspond to matches or non-matches based on other connections in the relationship graph. This is generally accomplished through an iterative approach that updates the weights (or probabilities) on the connections that determine the matching outcomes. Note that while in Fig. 6.8 only one type of connection needs to be classified, in the most general case not just connections between different types of entities, but also different types of connections, are available in a relationship graph.

The main differences between the various collective classification techniques are (1) how the relationship graph is generated from the underlying database(s) and (2) how the iterative update of the probabilities or weights in the graph, and their classification into matches (i.e. a connection exists between two nodes) or non-

AuthorID	Author name	Affiliation
a1	Dave Smith	Purdue
a2	Don Smith	Patras
a3	Susan Miles	Stanford
a4	John Black	Stanford
a5	Joe Green	?
a6	Liz Redman	?

PaperID	Co-author names
p1	John Black, Don Smith
p2	Susan Miles, **D Smith**
p3	Dave Smith
p4	Don Smith, Joe Green
p5	Joe Green, Liz Redman
p6	Liz Redman, **D Smith**

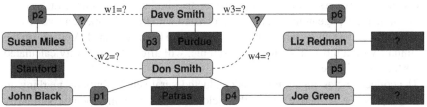

Fig. 6.8 Example of a graph-based collective matching approach of bibliographic records, adapted from [155]. The task is to identify (disambiguate) if the author 'D Smith' in papers 'p2' and 'p6' refers to either 'Don Smith' or 'Dave Smith'. Given Don Smith has co-authored paper 'p1' with 'John Black', who is affiliated with 'Stanford', and 'Susan Miles' is also affiliated with 'Stanford', there is a higher likelihood that 'Don Smith' rather than 'Dave Smith' is a co-author of paper 'p2', because 'Dave Smith' does not have any other connection with 'Stanford'. Similarly, given 'Don Smith' has written paper 'p4' with 'Joe Green', who has co-authored paper 'p5' with 'Liz Redman', there is a higher likelihood that 'Don Smith' is also the second co-author of paper 'p6' rather than 'Dave Smith' who has no connection to 'Joe Green'

matches (no connection exists between two nodes) is conducted. In the following, the major approaches to collective classification techniques are described in more detail.

- Kalashnikov and Mehrotra [155] build a relationship graph between different types of entities and with different relations, as the example graph in Fig. 6.8 shows. The disambiguation of connections between entities, i.e. their classification as being matches or non-matches, is conducted in an iterative approach where the unknown weights (such as 'w1' to 'w4' in Fig. 6.8) are updated based on the number of connections in the path that needs to be covered to get from one end of the connection under question to the other. For example, in Fig. 6.8, the path from 'Don Smith' via connection 'w4' continues onto 'p6', 'Liz Redman', 'p5', 'Joe Green', 'p4' and then back to 'Don Smith'. On the other hand, the only path starting from 'Dave Smith' would go via all other authors and even another path with unknown weight ('w1'), which is much less likely than the first path because it is a much longer path. Therefore, the weight for 'w4' can be set to a higher value than the weight for 'w3'. Kalashnikov and Mehrotra formalise this principle as the *context attraction principle* [155], and using this principle the unknown connection weights in the relationship graph are updated in an iterative fashion. An experimental evaluation on bibliographic data confirmed that this approach can lead to more accurate matching results compared to a pair-wise classification approach [155].

- Dong et al. [93] tackle the problem of collective classification of entities from multiple classes (types) by generating a dependency graph rather than a relationship graph. A node in the graph represents the similarity between a pair of entities of the same type, and a connection between nodes occurs when this similarity depends upon the similarity of another pair of entities. For example, the similarity between two papers (articles) depends upon the similarity between the titles, years of publication, page numbers, the authors listed with the two papers, as well as the similarity of the venues where the two papers have been published. A change in the similarity of authors or venues, for example, will affect the similarity calculated for the pair of papers.

  The collective classification task is conducted iteratively by initially marking all nodes as *active*. An active node is then selected, and depending upon the similarity in that node, it is either marked as *merged* (if its similarity is above a certain similarity threshold) or as *inactive* (otherwise). All neighbours of this just processed node that have a similarity below 1.0 (i.e. which do not have exact similarity) are then set as *active*. A queue of active nodes is maintained throughout the process, and in each iteration the similarity of the node at the top of the queue is recalculated. This process continues until no active node is left in the queue and all nodes are either marked as merged or inactive. This approach outperformed pair-wise classification techniques in experiments using several data sets [93].

- A machine learning based technique to collective classification has been proposed by Bhattacharya and Getoor [31]. In this approach, a relationship graph is built where the records (viewed as references to entities) are the nodes, and edges connect nodes if there is a relationship between them. For example, similar as shown in Fig. 6.8, the names of authors will be nodes in a reference graph, and all co-authors of a paper will be connected through an edge. These edges can be between more than two nodes, in which case they are called *hyper-edges*. If, for example, a paper was written by three co-authors, then one edge connects the three nodes that correspond to these co-authors. The similarity between two nodes is calculated as the weighted sum between the attribute value similarity and the relational similarity, where the latter considers the connectivity of two nodes through their hyper-edge as well as the connectivity of the neighbouring nodes they are connected to. Different relational similarity measures have been investigated [31].

  The collective classification is conducted using a priority queue that contains tuples made of two cluster identifiers and the similarity between the two clusters, sorted according to highest similarities first. An iterative algorithm merges clusters and updates the similarities between newly formed clusters as long as there are pairs of clusters in the queue that have a certain minimum similarity. When two nodes or clusters are merged, the similarities between the newly formed cluster and all its neighbours in the relational graph are updated, and the similarities between older clusters and the new cluster are added into the priority queue. The algorithm stops when no more clusters can be merged because the similarity between them is below the minimum threshold set by the user. Experiments on three different bibliographic databases showed that this approach is superior to pair-wise classification, however, at the cost of longer run times [31].

A variation of this relational clustering approach has been developed by Bhattacharya and Getoor to allow query-time collective classification [32]. A single query record is matched to a database that contains entity records and that can include duplicates. Using a collective classification approach, the query record is matched with the full database. While the reported matching accuracy of this approach is again very high, the matching time for a single query record was reported as being around 30 s, making this approach not suitable for real-time data matching (a topic that will be covered in detail in Sect. 9.3).

While collective classification techniques have shown to result in improved matching quality compared to pair-wise classification techniques, these improvements come at the cost of a higher computation complexity and thus reduced scalability to large databases. Recent work has aimed to improve the scalability of collective classification techniques by running a collective matching process many times on small subsets of records that are in the same neighbourhood of the data [225]. These independent collective matching instances exchange messages about the local matches found, and the results of all matching instances are combined into a final overall solution.

Thus far collective classification techniques for data matching have mostly been applied on databases that contain bibliographic data, or other data that contain several types of entities. It is not clear if and how collective classification techniques can be applied on data that only contain one type of entities, such as databases containing records about individuals.

6.11 Matching Restrictions and Group Linking

The classification of pairs or groups of records into the class of matches and non-matches discussed so far has not taken into account that in certain data matching applications there are restrictions with regard to the number of matches a single record can be involved in. The three possible scenarios when matching two databases, **A** and **B**, are:

- **One-to-one**: A record from **A** can match at most one record from **B**.
- **One-to-many**: A record from **A** can match at most one record from **B**, while a record from **B** can be involved in none, one or several matches with records from **A**. The one-to-many scenario is symmetric by swapping the databases **A** and **B**.
- **Many-to-many**: A record from **A** can match none, one, or several records from **B**, and a record from **B** can match none, one, or several records from **A**.

A one-to-one matching restriction is, for example, required when records from (historical) census databases are matched across time, and each record corresponds

Candidate pair	SimSum
(a1,b2)	4.5
(a1,b4)	5.5
(a2,b3)	5.9
(a2,b5)	4.9
(a3,b2)	5.8
(a3,b4)	4.7
(a4,b1)	5.3
(a4,b5)	6.0
(a5,b5)	5.1

Fig. 6.9 Examples of two approaches to enforcing a one-to-one assignment of matched candidate record pairs. The thicker lines between records illustrate the matched (assigned) records. The 'Optimal' approach aims to maximise the overall sum of the similarities (SimSum) over all matched record pairs, while the 'Greedy' approach matches candidate record pairs starting from the pair that has the highest similarity value until no more un-assigned records can be matched. In this example, the sum of similarity values for the matched (assigned) record pairs with the optimal approach is 27.6 while for the greedy approach the sum is only 23.2

to one individual [115, 116]. Because it is assumed that each census database only contains a single record per individual, one record in one census database (for example from 1900) can only match at most one record from another census database (for example from 1910).

A one-to-many matching restriction could be appropriate in a scenario where a client database of a government agency (that only contains one record per client) is updated with a set of new records that refer to individuals who in the recent past have been in contact with this agency. This new set of records can potentially contain several records for an individual because there might be several contact points for this government agency (online, telephone and face-to-face), and because this agency provides several programs (like a social security agency that provides housing, disability, unemployment and childcare support programs). Therefore, one client record in the cleaned and deduplicated client database maintained by this government agency can potentially match with several records in the set of new records.

A many-to-many matching is, for example, appropriate when two bibliographic databases are matched with the aim to identify and match all publications that refer to the same author, and there can be several records in each database that correspond to publications by one author. Returning to the example of matching census data, when the objective is to match households or families across census databases, rather than individuals, then a many-to-many matching scenario needs to be followed [114, 115].

While the clustering-based and collective classification techniques discussed in the previous two sections are mostly aimed at the many-to-many matching scenarios, the classification techniques presented in Sects. 6.2–6.7 classify individual record pairs independently from all others. Any one-to-one or one-to-many matching restriction can then be applied as a post-classification step on the set of candidate record pairs that were classified as matches.

A one-to-one matching restriction corresponds to finding an optimal solution to the problem of assigning individual records from the two databases into pairs (with one record originating in each database) based on the classified matched record pairs, such that the number of confirmed matched pairs and the sum of their similarities are maximised. As Fig. 6.9 illustrates, solving this problem corresponds to finding a solution to the *maximum weighted bipartite graph matching problem* [273].

A simple if not optimal approach to one-to-one matching is to sort the matched candidate record pairs according to their similarity values, and to assign pairs into the set of confirmed matches in a greedy fashion, as shown in the left graph in Fig. 6.9. The record pair with the highest similarity value is confirmed as a match first, and the two records of that pair are marked as being assigned matches (thick line). They can therefore not be part of any other matching pair. Then the next record pair (where both records are unassigned) with the highest similarity is confirmed as a match, and its two records are assigned as matches. This process is repeated as long as there are unassigned records that can be assigned to a record pair. For example, the pair (a4, b5) in Fig. 6.9 has the highest similarity value, SimSum = 6.0, and is therefore assigned as a confirmed match first. This, however, means that neither record 'a4' nor record 'b5' can be part of any other assigned pair. While this is a simple and fast approach (only requiring sorting the matching record pairs according to their similarities followed by a linear scan through that sorted list), this greedy approach is unlikely to produce a good solution because it is likely that not all records can be assigned into matching pairs. In Fig. 6.9, for example, the greedy approach cannot assign records 'a5' and 'b1' into a matching record pair.

Finding an optimal solution to the problem of assigning records into matching pairs is known as solving the *assignment problem*. Various algorithms have been developed to solve this problem [273]. One early approach is the so-called *Hungarian algorithm*, while another class of algorithms can solve this problem by viewing it as an auction problem [30]. The objective of an auction algorithm is to assign a group of people who all bid for several objects such that overall the highest profit can be obtained. People have maximum prices they are willing to pay for certain objects. When such an auction problem is mapped to the one-to-one matching restriction problem, people correspond to the records from one database, objects to the records from the second database, and the maximum prices to the similarities between pairs of records. Assignment algorithms are computationally more costly than the simply greedy approach presented before. Specifically, an auction algorithm has a computation complexity of the order $O(m \times n)$, where m is the number of links between records and n is the number of records involved [204, 205]. When a one-to-one matching restriction is required in data matching, then each subset of connected record pairs can be solved independently from all other subsets using an assignment algorithm applied on this subset only.

In some applications where many-to-many matchings are permissible, the main aim of a matching exercise is to identify groups of records that match across two databases rather than individual records [204, 205]. Groups can be defined according to some criteria, such as the value of a group identifier attribute. Example applications where such group linkage techniques are useful include the matching of families and households between census databases collected at different points in time [114, 115, 116], or the matching of bibliographic databases where sets of records correspond to the publications of one author [205]. The objective of group linkage is to identify an optimal matching of groups of records across two databases based on similarities calculated between individual pairs of records as well as a similarity measure that can be calculated for groups of individual record pairs. Both the Jaccard coefficient and a weighted bipartite graph matching approach have been successfully employed for the group linkage problem [204, 205].

6.12 Merging Matches

Thus far, it was assumed that the data matching process is completed once pairs or groups of records have been classified into matches and non-matches (with an acceptable quality as will be discussed in Chap. 7). In certain data matching and deduplication situations, however, matched records also need to be merged (in some way) before the matched data can be used further, either for data analysis or data mining, or for further data processing such as generating mailing lists. In this last example, the objective of a data matching exercise is to create a database that contains complete, accurate and up-to-date address and name details for all records in a mailing list. Achieving this goal means that the values in certain attributes for the matched records need to be merged.

While traditionally the merging of matched records has not been considered by most research in data matching, a recent research project has investigated how this merging step can be best incorporated into the overall data matching process. The Stanford Entity Resolution Framework (SERF) project [25, 26, 186] has developed generic data matching techniques that assume the actual matching of records as a black-box approach, represented as a function $match(r_i, r_j)$, which returns true if two records are matching and false otherwise. An additional black-box function, $merge(r_i, r_j)$, is defined on matching record pairs. It returns a new record that is generated by (somehow) merging the content of records r_i and r_j. While the actual merge function is domain and application specific, a *merge domination* is defined as the situation when for two records r_i and r_j it holds $merge(r_i, r_j) = r_j$. When the merge function corresponds to combining attribute values from r_i and r_j, r_j dominating r_i means that r_i does not contribute any new attribute value(s) to the merged record beyond what r_j already contains.

The generic entity resolution process on a database consists of an iterative matching and merging approach which results in a set of merged records that cannot be further matched or merged with each other, and no merged record is dominated by another merged record [26]. Based on these assumptions, a set of entity resolution algorithms (named G-Swoosh, R-Swoosh, F-Swoosh, D-Swoosh, and P-Swoosh) were developed by the SERF project. The G-Swoosh algorithm has no particular requirements on the *match* and *merge* functions. It helps to illustrate the process of entity resolution. In the R-Swoosh algorithm, if two matched records r_i and r_j

are merged into $r_{i,j}$, i.e. $r_{i,j} = merge(r_i, r_j)$, then the new record $r_{i,j}$ is added into the set of all records and the two original records r_i and r_j are removed from this set. This approach also means that dominated records do not need to be explicitly removed from the set of all records as they are eliminated in the merge and removal step.

The F-Swoosh algorithm improves performance by taking feature (attribute value) comparisons into account such that each pair of features is only compared once. D-Swoosh [25] and P-Swoosh [160] are algorithms aimed at distributed and parallel computing environments, respectively. Both these algorithms are described further in Sect. 9.5.

A more recently proposed approach is to employ locality sensitive hashing (LSH) for quick iterative blocking of the records in a databases [164]. All records hashed into the same bucket (block) by the hash-algorithm are matched and merged, and the merged records are re-hashed. This process is repeated until either no more matches and merges are found, the reduced number of record pairs reaches a certain minimum number, or a specified maximum number of iterations has been reached. The authors proposed several variations of their approach depending upon if the databases to be matched contain duplicates or not. Experimental results on a bibliographic database showed that this hash-based approach is able to achieve better scalability to large databases compared to the R-Swoosh algorithm [164].

6.13 Practical Considerations and Research Issues

The choice of what type of classification technique to employ for a certain data matching or deduplication exercise depends upon various factors, including the classification techniques available in the matching software that is used (or the techniques that can be implemented), and the type of data that are to be matched or deduplicated. If a supervised classification technique is to be used, training data in the form of record pairs with their known match status are needed.

A suggested approach is to evaluate different classification techniques, and in the case where no training data are available, to manually generate a set of record pairs (together with their match status) that represent the characteristics of the data (such as the distribution of values, and the types and distribution of errors and variations in the data that are to be matched). While time-consuming and labour-intensive, such an approach will enable an evaluation and comparison of the classification accuracies of different data matching algorithms.

Unfortunately, no comprehensive survey of classification techniques for data matching and deduplication has so far been published. What is needed is an experimental evaluation of different techniques on a variety of test data sets from different domains and of different sizes. These data sets should contain the true match status of record pairs so that the resulting matching quality can be evaluated. Data sets of different sizes are required so that the scalability with regard to training time, classification time and memory usage can be evaluated.

Future research in the area of classification for data matching and deduplication should be aimed at investigating if and how collective classification techniques can be applied to data that do not contain different types of entities (for example, data containing personal details such as names and addresses), and how classification techniques can be employed on very large databases that contain many millions of records. Given the difficulties of obtaining or generating training data (as will be discussed further in Chap. 7), a major focus of research should be on unsupervised and automatic classification techniques that do not require manual preparation of training data.

Another area of future research is the development of adaptive classification techniques, given that in many application areas data matching is no longer employed in batch mode and on static databases. Rather, in many modern information systems data matching and deduplication functionalities are integrated into larger systems where new records that contain the details of entities are being added into databases or data warehouses in an ongoing basis. Matching in real time and matching dynamic databases will be discussed further in Sects. 9.3 and 9.4.

6.14 Further Reading

The book by Herzog, Scheuren and Winkler [143] contains arguably the most accessible and detailed description of the probabilistic record linkage approach. Issues such as the conditional independence assumption and parameter estimation are discussed in detail and illustrated via examples. Further examples of probabilistic record linkage applications are also provided. Talburt nicely explains the Swoosh-based entity resolution approaches using several small example databases [249]. He also describes an algebraic model for data matching. For general introductions to classification techniques, the reader is referred to textbooks in the areas of machine learning or data mining [135, 189].

Naumann and Herschel [195] cover graph-based and collective classification techniques, as well as clustering and rule-based approaches (even though in their book rules-based approaches are discussed under the topic of comparison functions). Batini and Scannapieco [19] also provide an overview of different techniques, including a brief comparison with regard to the requirements (such as expected input, generated output and classification objectives with regard to a quality metric) of different classification techniques for data matching.

The best coverage of the topic of how to merge pairs or groups of records that have been classified as matches is provided by Benjelloun et al. [25] in their description of the techniques developed in the SERF project. Data fusion more generally is covered in the recent survey by Bleiholder and Naumann [38]. For a tutorial on the assignment problem that can be employed to finding a solution to the one-to-one matching problem the reader is referred to the excellent tutorial provided by Bertsekas [30].

Chapter 7
Evaluation of Matching Quality and Complexity

7.1 Overview

Over the past decades, as the previous chapter has shown, various classification techniques for data matching have been developed. The main objective of these techniques is to achieve high matching quality. Similar to other classification problems, in order to be able to assess the quality of the matched data for a certain data matching project, ground-truth data, also known as 'gold standard' data, are required. The characteristics of such ground-truth data must be as close as possible to the characteristics of the data that are to be matched.

To summarise, if a record pair has been classified as a match, then the assumption is that both records in the pair refer to the same real-world entity. For a record pair classified as a non-match, on the other hand, the two records in the pair are assumed to refer to two different real-world entities. Thus if a ground-truth data set with known true matching and non-matching record pairs is available, then similar to other classification problems in machine learning and data mining [135], a variety of measures can be calculated on the outcomes of the classification process. Several such measures are discussed in this chapter.

The question now arises: how to acquire such ground-truth data for a certain data matching exercise. In many if not most data matching situations no ground-truth data are readily available. There are several approaches of how ground-truth data can be generated.

One possibility is that the results from a previous data matching project in the same domain (ideally an earlier version of the same databases) are available, and that these databases have been manually evaluated with regard to the quality of the previous matching outcomes. For example, domain experts might have detected wrongly matched as well as missed true matching pairs of records as they have worked with the matched databases. The quality of previously matched data might however not be good enough to be used as training data, especially if a more simpler matching approach was previously employed. Additionally, the manual inspection and possible correction of matches are often not 100 % correct, and it is therefore

P. Christen, *Data Matching*, Data-Centric Systems and Applications,
DOI: 10.1007/978-3-642-31164-2_7, © Springer-Verlag Berlin Heidelberg 2012

likely that the databases used as ground-truth data contain mistakes with regard to the match status of certain record pairs.

Another approach to obtain ground-truth data is to manually generate such data by sampling pairs of records from the two databases that are to be matched (or pairs from the single database that is to be deduplicated), and to manually classify these pairs as being either a match or a non-match. This approach has two difficulties.

The first is similar to the drawbacks described above when a previously matched database is used as ground-truth data, in that the manual classification of record pairs is unlikely to always be correct. Some mistakes will potentially be introduced by a human classification. These mistakes will not be in the record pairs that are easy to classify. Two records that differ in all attribute values are very obviously not a match. Similarly, two records where all attribute values are the same or only contain minor differences can be manually classified as a match with high confidence. These two types of record pairs can also be easily classified automatically, as was discussed in the previous chapter.

However, the record pairs that contain variations or differences in several of their attribute values are hard to classify. These variations include ambiguous names, or changed name or address values that are due to a person having married or moved to a new address. Often, additional information is required so an accurate manual classification can be performed. Section 7.4 will further cover this topic in the context of manual clerical review of potential matches.

The second drawback when manually generating ground-truth data based on sampling record pairs from the databases to be matched is the overall distribution of matches and non-matches in the classified record pairs. Assuming no indexing or blocking (as discussed in Chap. 4) has been applied, the matching of two databases that contain m and n records, respectively, will generate $m \times n$ record pairs that need to be classified. If it is assumed that both databases have been deduplicated prior to the matching, then a maximum of $\min(m, n)$ true matching record pairs are contained in the $m \times n$ record pairs. The number of true matches is therefore much smaller than the number of non-matches, especially as the size of the databases increases. The same holds for the deduplication of a single database that contains m records, where (without indexing) $m(m - 1)/2$ record pairs need to be compared, but where the maximum number of true matches will be $m - 1$ (in the unlikely case where $m - 1$ records are duplicates of one single record). The sizes of the match and non-match classes are therefore often very imbalanced in data matching. Even if some form of indexing has been applied, the number of candidate record pairs generated is very likely to be much larger than the number of true matches contained in them.

Using simple random sampling of candidate record pairs and manually classifying the sampled pairs to generate a ground-truth data set will therefore result in a sampled set that mostly contains non-matching record pairs. A stratified sampling approach can be employed, such that a balanced number of true matches and non-matches is sampled from all record pairs. This can for example be achieved by binning the comparison vectors of record pairs according to their summed similarities, and then sample the same number of comparison vectors from each bin.

A third approach to obtain data that contain the true match status of record pairs is to use one of the small number of publicly available test data sets that have been generated by researchers to test their algorithms. An overview of such data sets is given later in this chapter in Sect. 7.5. Finally, a fourth approach is to use synthetically generated data that have similar characteristics as the real databases that are to be matched. This approach will be discussed further in Sect. 7.6. For these last two approaches, the match status of record pairs is generally known. If it makes sense to use public or synthetic data sets to evaluate a certain data matching system in practice depends upon the actual situation in which a data matching system will be employed.

7.2 Measuring Matching Quality

Assuming some ground-truth data sets with the true match status of all its possible record pairs is available and a matching has been conducted on these data sets, each compared and classified record pair is assigned into one of the following four categories [71]:

- *True positives*. These are the record pairs that have been classified as matches and that are true matches. These are the pairs where both records refer to the same entity.
- *False positives*. These are the record pairs that have been classified as matches, but they are not true matches. The two records in these pairs refer to two different entities. The classifier has made a wrong decision with these record pairs. These pairs are also known as false matches.
- *True negatives*. These are the record pairs that have been classified as non-matches, and they are true non-matches. The two records in pairs in this category do refer to two different real-world entities.
- *False negatives*. These are the record pairs that have been classified as non-matches, but they are actually true matches. The two records in these pairs refer to the same entity. The classifier has made a wrong decision with these record pairs. These pairs are also known as false non-matches.

Figure 7.1 illustrates these four outcomes. The true positives are the intersection of the true matches and classified matches. It is common to illustrate these four possible outcomes of a classification in a confusion or error matrix [135], as shown in Fig. 7.2.

As was discussed previously, the number of true negatives in data matching situations will often be much larger than the sum of the number of true positives, false negatives and false positives. The reason for this is the nature of the comparison process, because there are many more pairs where the two records refer to two different entities than there are pairs where both records refer to the same entity [71].

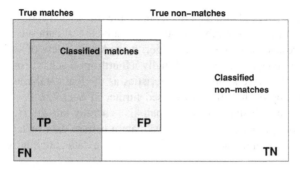

Fig. 7.1 Example illustration of the classification outcomes for data matching. *TP* refers to true positives, *FP* to false positives, *TN* to true negatives and *FN* to false negatives, as discussed in Sect. 7.2

		Predicted classes	
		Matches	Non-matches
Actual	Matches	True Positives (true matches)	False Negatives (false non-matches)
classes	Non-Matches	False Positives (false matches)	True Negatives (true non-matches)

Fig. 7.2 Error or confusion matrix illustrating the outcomes of a data matching classification. As discussed in Sect. 7.2, the aim of the classification is to identify and correctly classify as many true matches as possible while keeping the number of false positives and false negatives as small as possible

An ideal outcome of a data matching project is to correctly classify as many of the true matches as true positives, while keeping both the number of false positives and false negatives small.

Based on the number of true positives (TP), true negatives (TN), false positives (FP) and false negatives (FN), different quality measures can be calculated [71]. The following list presents the measures most commonly used with data matching, and discusses their characteristics and their suitability for assessing the quality of data matching and deduplication.

- *Accuracy*. This quality measure is calculated as

$$acc = \frac{TP + TN}{TP + FP + TN + FN}. \tag{7.1}$$

This measure is most widely used for binary as well as multi-class problems in the fields of machine learning and data mining [135]. Accuracy is mainly useful for situations where the classes are balanced, i.e. where the number of instances (record pairs in the case of data matching) are more or less the same for both classes (matches and non-matches).

As was previously discussed, balanced classes are rare in data matching and dedu-plication classification, in that the majority of record pairs corresponds to true non-matches (true negatives). As a result, the accuracy measure is not suitable to properly assess matching quality. The value of TN in Eq. 7.1 dominates the calculation of accuracy.

For example, assume two databases with $1,000,000$ records each are matched, and the indexing step resulted in $50,000,000$ candidate record pairs that were gener-ated. Now assume that there are $500,000$ true matches between these two databases. Also assume a classifier has classified $600,000$ record pairs as matches, and of these $400,000$ correspond to true matches. As a result, there will be TP $= 400,000$, FN $= 100,000$, FP $= 200,000$ and TN $= 49,300,000$. The accuracy calculated on these values then is: acc $= \frac{400,000+49,300,000}{50,000,000} = 0.994$, which corresponds to 99.4% accuracy. Clearly, this is not a meaningful measure, because only $400,000$ of the $500,000$ true matches were classified correctly. Even a simple classification of all candidate record pairs as non-matches (TP $= 0$, FN $= 500,000$, and FP $= 0$) will still result in a very high accuracy value.

As a result, because of the imbalanced classification problem that data matching and duplication commonly pose, accuracy is not a suitable quality measure and should not be used. The following measures are more appropriate alternatives.

- *Precision*. This is a measure commonly used in information retrieval to assess the quality of search results [288]. It is calculated as

$$\text{prec} = \frac{\text{TP}}{\text{TP} + \text{FP}}. \tag{7.2}$$

Because precision does not include the number of true negatives, it does not suffer from the class imbalance problem in the way accuracy does. Precision calculates the proportion of how many of the classified matches (TP + FP) have been correctly classified as true matches (TP). It thus measures how precise a classifier is in classifying true matches. Precision is also known as the positive predictive value (PPV) in the medical literature [37].

Using the same numerical values as in the example above, precision can be cal-culated as: prec $= \frac{400,000}{400,000+200,000} = 0.667$, which corresponds to a precision of 66.7%. This means that two-thirds of the record pairs this classifier has classified as matches correspond to true matches, while a third corresponds to false matches.

- *Recall*. This is a second measure that is commonly used in information retrieval [288]. It is calculated as

$$\text{rec} = \frac{\text{TP}}{\text{TP} + \text{FN}}. \tag{7.3}$$

Similar to precision, because recall does not include the number of true negatives, this measure does not suffer from the class imbalance problem. It measures the proportion of true matches (TP + FN) that have been classified correctly (TP). It thus measures how many of the actual true matching record pairs have been correctly classified as matches. Recall is also known as the *true positive rate* or the *hit rate*, while in the medical literature it is known as *sensitivity* [301] and commonly used to assess the results of epidemiological studies.

Continuing the numerical example, the recall of this classification outcome can be calculated as: $rec = \frac{400,000}{400,000+100,000} = 0.8$, which corresponds to a recall of 80.0 %. This means that this classifier has correctly classified four out of every five true matching record pairs.

It should be noted that there is a trade-off between precision and recall. Depending upon the data matching or deduplication situation, it might be more important to achieve matching results with high precision but accept a lower recall, while in other data matching situations having a low precision is acceptable but a high recall is required.

For example, for a crime investigation where certain suspect individuals need to be matched with a large database of people, a high recall is desired to make sure that it is likely that the individuals one is looking for are included in the matched record pairs, even if there is a larger number of matches that need to be investigated. On the other hand, high precision is required in many public health studies where each match would correspond, for example, to a patient with certain medical characteristics who needs to be included into a cohort study. In this situation one wants to be sure to only include patients into the set of matched record pairs who do have the medical condition one is interested in.

- *F-measure.* This measure, also known as *f-score* or f_1-*score*, calculates the harmonic mean between precision and recall [195]:

$$fmeas = 2 \times \left(\frac{prec \times rec}{prec + rec} \right). \tag{7.4}$$

The *f-measure* combines precision and recall and only has a high value if both precision and recall are high. Aiming to achieve a high *f-measure* requires to find the best compromise between precision and recall [15].

With the continuing numerical example, with $prec = 0.667$ and $rec = 0.8$, then $fmeas = 2 \times (\frac{0.667 \times 0.8}{0.667+0.8}) = 0.727$.

- *Specificity.* This measure is also known as the *true negative rate* and it is commonly used in the medical literature [301]. It is calculated as

$$spec = \frac{TN}{TN + FP}. \tag{7.5}$$

Because this measure includes the number of true negatives (TN), it suffers from the same problem as accuracy, and should not be used for data matching or deduplication. If the number of false positives (FP) is small compared to the number of TN (which it likely is because of the class imbalance in data matching), then the calculated specificity will be dominated by the number of TN.

For the numerical example, specificity is calculated as $spec = \frac{49,300,000}{49,300,000+200,000} = 0.996$, which corresponds to 99.6 %.

- *False positive rate*. This measure is also known as *fall-out* in information retrieval [288]. It is measured as

$$\text{fpr} = \frac{\text{FP}}{\text{TN} + \text{FP}}. \tag{7.6}$$

Note that fpr = $(1 - \text{spec})$. Because this measure includes the number of true negatives (TN), it suffers from the same problem as accuracy and specificity, and should not be used for data matching.

Continuing the numerical example, the false positive rate is calculated as fpr = $\frac{200,000}{49,300,000+200,000}$ = 0.004, which is a very low 0.4 %. This very low value does not reflect that this classifier classified 80 % of the true matches correctly.

The measure most commonly used in the computer science literature for assessing the quality of data matching has been accuracy [102, 129, 155, 231, 252, 301]. However, precision [29, 83, 185], recall [129, 185, 301] and the f-measure [31, 85, 185] have also been used, and they have gained popularity in recent years as researchers have become more aware of the pitfalls of using the accuracy measure [71].

While the above measures provide a single number of the matching quality achieved by a single classifier, most classification techniques described in Chap. 6 have one or several parameters that can be modified and tuned. Depending upon the value(s) of such parameter(s), a classifier will have a different performance, leading to different numbers of true and false positives and negatives. Rather than a single value for a quality measure using a certain parameter setting, a series of values can be generated for a certain measure using different parameter settings. The resulting values can then be visualised in various ways to illustrate the performance of a certain classifier over a range of parameter settings. Such visualisations also allow more detailed comparisons of the performance of several classification techniques. The following three visualisations (also shown in the example in Fig. 7.3) are commonly used to illustrate the outcomes of the classification of candidate record pairs.

- *Precision–recall graph*. In this visualisation the values of precision and recall are plotted against each other as generated by a classifier with different parameter settings. This type of graph is commonly used in the field of information retrieval to visualise, for example, the quality of results returned by a Web search engine [288]. Figure 7.3b shows an example of a precision–recall graph.

  For each selected parameter setting of a classification model, the precision and recall values are calculated resulting in a single point in the precision–recall graph. Recall is plotted along the horizontal axis (or x-axis) of the graph, while precision is plotted against the vertical axis (or y-axis). As parameter values are changed, the resulting precision and recall values generally change as well.

  Commonly, a high precision of a classifier will result in a low recall value and vice versa. Therefore, in precision–recall graphs there is often a curve starting in the upper left corner moving down to the lower right corner. Ideally, a classifier should achieve both high recall and high precision and therefore the curve should be as high up in the upper right corner as possible.

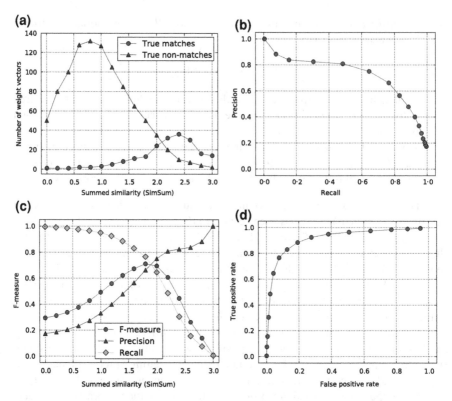

Fig. 7.3 An example of simulated classification results assuming 200 true matching and 1000 true non-matching record pairs, and three possible visualisations of the quality results of such a classification. Plot **a** shows the distribution of the comparison vectors summed into similarity values SimSum (as was previously illustrated in Fig. 2.6 on page 31). As discussed in this chapter, the majority of the compared record pairs refer to true non-matches. Plots **b** to **d** show the different quality graphs that can be generated from the summed similarities in Plot **a**. **a** Summed Similarities **b**. Precision-Recall **c**. F-measure **d**. ROC curve

- *F-measure graph.* An alternative to plotting two quality measures (such as precision and recall) against each other is to plot the values of one or several measures with regard to the setting of a certain parameter, for example a single threshold used to classify candidate record pairs according to their summed comparison vectors, as was discussed in Sect. 6.2. This is shown in Fig. 7.3c. In this graph, the horizontal axis shows the summed similarity score (SimSum) that is used as classification threshold. For each threshold value, all record pairs with a summed similarity below that threshold will be classified as non-matches and all other pairs as matches. As the threshold is increased from 0 to 3, for this example, the number of record pairs classified as non-matches increases (and thus the number of TN and FN increases), while the number of TP and FP decreases.

Any of the above discussed quality measures can be plotted in such a graph. An often used combination is to show precision, recall and the f-measure in the same graph, as illustrated in Fig. 7.3c. As the classification threshold is increased, the value of recall gets lower (because less of the true matches, those with a lower overall similarity, are classified as matches), while precision gets higher (because less true non-matches are classified as matches with higher similarity threshold).

- *ROC curve.* Similar to the precision–recall graph, the receiver operating characteristic (ROC) curve is plotted as the values of two quality measures against each other [106]. The horizontal axis is the false positive rate while the vertical axis is the true positive rate (which is the recall). The closer an ROC curve is to the top left corner the better a classifier is, because this means it can achieve a high recall with a small number of false positives.

While the use of ROC curves is being promoted to be robust against imbalanced classes (as is common in data matching and deduplication) [106], the problem when applying them in data matching is that the number of true negatives, which is a factor only when the false positive rate is calculated, will lead to too optimistic ROC curves because the false positive rate will be calculated to be very low. The use of ROC curves for data matching should therefore be carefully assessed. Plotting several ROC curves generated by different classifiers can certainly help to compare their performance over a range of parameter settings.

Based on an ROC curve, a numerical measure called the area under the curve (AUC) can be calculated. This is basically the area of an ROC graph that lies in the lower right area of the graph below the curve. The closer an ROC curve is to the upper left corner the larger its AUC value becomes, and therefore the better a classifier performs over a range of parameter values. Note that the value of AUC is always between $0.5 \le auc \le 1.0$, because even a random classifier (which would have an ROC curve that is the diagonal in the ROC graph) has an AUC value of $auc = 0.5$, while a perfect classifier will have an AUC of $auc = 1.0$.

The issue of how to evaluate the merging of matched records into new compound records has recently been investigated by Menestrina et al. [187]. In their work, the authors compare different measures that have been used by researchers in data matching. They show that assessing the outcomes of a data matching project using different measures can lead to different rankings of the matched record pairs, and thus to different matching outcomes. The merging of records into entities is viewed as a clustering process. Each clustering result is compared to the ground-truth data (gold standard). A generalised merge distance (GMD), defined by the authors, is used to assess how close a certain clustering is to the known true clustering of records into entities (where each cluster refers to one entity). The GMD is related to the edit distance, as discussed in Sect. 5.3, in that it assigns costs to merging a cluster of records or splitting a cluster. The smaller the number of merges or splits is from a given set of clusters (the results of a data matching classification) to the true clustering result, the smaller the GMD is. The authors also show that precision, recall and thus the f-measure (Eqs. 7.2–7.4), can all be calculated easily from the GMD.

7.3 Measuring Matching Complexity

Besides the quality of the record pairs classified as matches and non-matches within a data matching or deduplication project, a second major aspect is the efficiency and effectiveness of data matching techniques or systems. One obvious approach would be to simply measure the run-time on different data sets to compare which technique or system is faster. The results of such an assessment would be very specific to the computing hardware used, such as the speed of its processors, and its memory and I/O bandwidths. A platform-independent way to compare systems or techniques would be of advantage because results would be more generalisable. One possibility is to count the number of candidate record pairs generated by an indexing technique, and use this number to calculate a measure of how complex a data matching exercise is. Three such approaches to measuring the efficiency and complexity of data matching and deduplication have been proposed [71].

Following the notation given in previous publications [20, 64, 71, 102, 128], the total number of matched and non-matched record pairs are denoted with n_M and n_N, respectively, with $n_M + n_N = m \times n$ for the matching of two databases that contain m and n records, respectively, and $n_M + n_N = m(m - 1)/2$ for the deduplication of one database that contains m records. Note that these numbers correspond to the full comparison space of all possible record pairs, i.e. when no indexing has been applied. The number of true matched and true non-matched candidate record pairs generated by an indexing technique is denoted with s_M and s_N, respectively, with $s_M + s_N \leq n_M + n_N$. Three measures can now be defined.

- *Reduction ratio.* This measure provides information about how many candidate record pairs were generated by an indexing technique compared to all possible record pairs, without assessing the quality of these candidate record pairs. Reduction ratio is calculated as

$$rr = 1 - \left(\frac{s_M + s_N}{n_M + n_N} \right). \tag{7.7}$$

The reduction ratio therefore measures the relative reduction of the comparison space of a data matching or deduplication exercise. A high reduction ratio means an indexing technique has removed many record pairs from the full comparison space, while a low reduction ratio means that a larger number of candidate record pairs have been generated.
- *Pairs completeness.* This measure takes the true match status of candidate record pairs into account. It is calculated as

$$pc = \frac{s_M}{n_M}. \tag{7.8}$$

Pairs completeness corresponds to recall (Eq. 7.3) discussed previously. It is the number of true matching record pairs that have been generated by an indexing technique divided by the total number of true matching pairs in the full comparison space. The lower the pairs completeness value is the more true matches have been removed by an indexing technique. This leads to lower matching quality, because the record pairs removed in the indexing step are implicitly classified as non-matches without being compared in detail.

There is a trade-off between reduction ratio and pairs completeness [20], i.e. between the number of record pairs that are removed in the indexing step and the number of missed true matches. No indexing technique is perfect in only removing record pairs that correspond to non-matches. Some record pairs that correspond to true matches are likely removed as well in the indexing step. Using an indexing technique that has a lower reduction ratio will mean that a smaller number of candidate pairs is removed in the indexing step. This can often lead to an increased pairs completeness value.

- *Pairs quality*. This third measure, which also takes the quality of candidate record pairs into account, is calculated as

$$pq = \frac{s_M}{s_M + s_N}. \tag{7.9}$$

It is the number of candidate record pairs that correspond to true matches that were generated by an indexing technique, divided by the total number of candidate record pairs that were generated. It corresponds to the measure precision (Eq. 7.2) presented previously. A high pairs quality value means that an indexing technique is successful in generating candidate record pairs that mostly correspond to true matches, while keeping the number of candidate pairs that correspond to non-matches low. Similar to the trade-off between precision and recall, there is normally a trade-off between pairs completeness and pairs quality. Aiming for an increase in one of these two measures generally results in a decrease in the value of the other measure.

None of these three measures is taking any computational resources into account, such as processing time or main memory usage. These are dependent upon the actual implementation of a data matching system and the computing platform used.

Similar to all quality measures discussed in Sect. 7.2 above, being able to calculate both the pairs completeness and pairs quality measures does require knowledge about the true match status of record pairs. If this information is not available for a given data matching or deduplication exercise, then only the reduction ratio, as well as run-time and memory usage, can be reported. This prohibits a proper assessment of a data matching technique or system.

7.4 Clerical Review

The traditional classification model of probabilistic record linkage (discussed in Sect. 6.3) that has been used in data matching for several decades (and that is implemented in various data matching systems) classifies the compared candidate record pairs into matches, non-matches, as well as potential matches, as Fig. 2.1 on page 24 shows. The class of potential matches consists of record pairs where a decision model was not able to make a clear decision on if they correspond to matches (where both records refer to the same entity) or non-matches (where the two records refer to two different entities) [129].

A manual classification is required for candidate record pairs that have been inserted into the class of potential matches. This manual classification requires a clerical review where each record pair is assessed visually and a match decision is made manually. Figure 7.4 provides an example that illustrates how a pair of records might be presented to a person who conducts such a manual review. Smith and Newcombe in an early study using health records showed that a computer-based probabilistic data matching system can result in more reliable, consistent and more cost effective matching results compared to a fully manual approach [241, 242].

The manual clerical review of potential matches can be a tedious, time-consuming and labour intensive process, especially in cases where the matching of two databases has resulted in a large number of record pairs that were classified as potential matches. This can either be because the databases were large, the data were difficult to classify, or the classification model was not able to accurately discriminate matches from non-matches.

Several aspects make manual clerical review a difficult process. First of all, looking at Fig. 7.4, one can see that even for an experienced domain and data matching expert it can be difficult to make an accurate manual classification when assessing a single record pair in isolation. Other records might be similar and have characteristics that also make them potential match candidates even though they have a lower overall similarity. For the given example, there might be another record with surname 'Stevens', given name 'Sal' and age '17' but with a missing gender value, and an overall similarity of 72 %. This could well be a better matching record. Ideally, therefore, a system that facilitates manual clerical review should visualise not just a pair of records but a whole group of similar records, for example in the same way as a Web search engine presents a list of query results ranked according to relevance. Alternatively, having access to external data that can help validate if a pair of records corresponds to a match or non-match can be highly beneficial in the manual decision-making process. Such external data can for example be a database that contains the known previous addresses or telephone numbers of individuals, or their known nicknames and previous surnames.

A second issue that makes clerical review a difficult process is that the manual match or non-match decision made can differ not only from reviewer to reviewer, but even the same reviewer might make different decisions depending upon their mood, time of day and concentration level. The same reviewer in the morning might classify

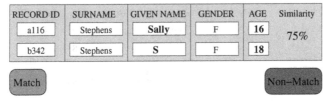

Fig. 7.4 Example for a clerical review of a record pair classified as a potential match with an overall similarity calculated to be 75 %. The fields (or attributes) where values are different are shown with a larger font to grab the attention of a reviewer. The 'Match' and 'Non-Match' buttons allow a reviewer to make their manual classification decision

a pair as a match, but if they would see the same pair late in a day's work might decide it is a non-match based on their mood and desire to finish a day's work. It is therefore of advantage to have more than one reviewer assessing the same set of record pairs, so that in case of a conflicting classification of a certain pair an additional review can be asked for. This of course prolongs the review process and also makes it more costly. As with many aspects of data matching and deduplication, the application of where the matched records will be used dictates how accurate the matched data need to be and how costly a false match or a false non-match will be.

One advantage of the clerical review process is that it can help generate training data of record pairs that are difficult to classify. Such manually generated training examples can flow back into the classification step as illustrated in Fig. 2.1 on page 24. It is however important to be aware of the above discussed issues, and that the class of potential matches does contain the most difficult to classify record pairs. The confidence one can have on the manually classified record pairs depends upon the thoroughness of the manual review process, the system used to present potentially matched pairs to the reviewer, the domain expertise of the reviewers, and if they had access to any external data that supported their decisions. Using manually classified pairs as training data for a record pair classifier therefore needs to be considered carefully.

One way to reduce the possibly large number of potential matches that need to be classified manually is to employ an active learning approach as was discussed in Sect. 6.7. With active learning, only a small number of hard to classify record pairs are manually classified in each iteration, and a new classifier is trained using training data that include these manually classified pairs. The same careful consideration as mentioned before about the confidence one has into the manually classified pairs needs to be considered when active learning techniques are employed. Ideally, an active learning classifier should be able to take a confidence level of the manually classified record pairs into account, as illustrated in Fig. 7.5.

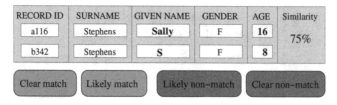

Fig. 7.5 A variation of Fig. 7.4 that allows a reviewer to provide feedback about the confidence of their manual classification decision

7.5 Public Test Data

As discussed in Sect. 7.2, knowing the true match status of record pairs is a requirement for being able to measure the matching quality of a data matching or deduplication system for a certain data set. For many real-world applications, however, it is very difficult to obtain or create such ground-truth data. Even if significant manual resources are put into a manual training process (which is similar to the manual clerical review process described in the previous section), then for many of the record pairs added to the training set the match status might not be known with high confidence.

For researchers, who might be working in an academic environment without close collaboration with an organisation that has real data and that is prepared to provide these data for research, it is generally very difficult to get access to any real-world data sets.

As an alternative, researchers have investigated public sources of data that can be used to test and evaluate data matching and deduplication techniques. Because of the private nature of personal information (such as names, addresses, date of birth and so on), data that contain information other than personal details are commonly used by researchers. The issues involved with privacy in data matching are covered further in Chap. 8.

Over the years, a collection of test data sets have been used by various researchers in the field of data matching. The most common of these data sets are described in the following list:

- *Cora.* This data set contains 1,295 bibliographic records of machine learning articles that correspond to 189 actual real publications. A total of 17,184 out of $1,295 \times 1,294/2$ pairs of records correspond to true matches (assuming no indexing is applied). Each record in this data set contains the publication name, the publication year, one or more author names (sometimes only surnames and initials) and the conference or journal name (or their abbreviation only) where an article was published.
- *Restaurant.* This small data set contains 864 records of restaurant names, addresses, telephone numbers and food style (French, Italian, Japanese, etc.) taken from the Fodor and Zagat restaurant guides. In total, 752 different restaurants are included in this data set, and there are 112 restaurants that appear twice.

- *Census*. This is a pair of small data sets that contain synthetic census records generated by the US Census Bureau. The first data set contains 449 records while the second data set contains 392 records. The number of matching records is 327. Each record consists of the attributes first and last name, middle initials, a street number and a street name.
- *UCD people*. This is a data set which contains the names of people working at the University College in Dublin. Each person is represented by a single string that contains the person's given name and surname, as well as optional titles and a role or position description such as 'Head of Department' or 'Newman Scholar'. Each individual is assigned a unique identifier.
- *CDDB*. This data set consists of 9,763 records with details of compact disc albums (CDs), such as their artist, title, genre and the year when the CD was published. These records were randomly selected from the FreeCD database [195]. A time-consuming manual process lead to the detection of 298 true duplicates in this data set (with a total of 607 true matching record pairs, assuming no indexing is applied). Each unique CD was given a unique identifier.
- *DBLP*. This online database containing computer science journals and conference and workshop proceedings with over a million articles has been used by several data matching research groups for their experimental studies. Each record consists of an article's name, details of the publication venue, its year of publication and the names of the author(s) of that publication. An XML version of this database[1] can be freely downloaded. A drawback of this database is however that the true match status is not known, and therefore it is difficult to use this database to evaluate the accuracy of data matching techniques without some initial processing and generation of some form of ground-truth data.
- *IMDB*. The Internet movie database[2] is another popular source of data used to evaluate data matching and deduplication techniques. The database contains details about different types of entities, such as people (actors, producers, directors, etc.), movies, companies, as well as movie ratings and plot descriptions. Similar to DBLP, no ground-truth data are available. However, compared to DBLP, where it is likely that duplicate records exist for the same article, in the IMDB database it is more likely that several records with the same name refer to different individuals (there are, for example, more than 20 people with the name 'Bill Murray' listed in IMDB), or that several movies have the same name. Researchers commonly corrupt the data they download from IMDB to generate duplicate records. This approach to artificially generating data is discussed further in the next section.

The Cora, UCD people, Restaurant and Census data sets are available in several repositories and open source data matching systems, including the RIDDLE repository[3] the SecondString toolkit,[4] and the FEBRL system.[5] Note that there are various

---

[1] http://dblp.uni-trier.de/xml/

[2] http://www.imdb.com/interfaces

[3] http://www.cs.utexas.edu/users/ml/riddle/

[4] http://secondstring.sourceforge.net

[5] http://sourceforge.net/projects/febrl/

versions of the Cora data set in different repositories. Some have been pre-processed and cleaned further than the original version. The CDDB and Cora data sets (as well as some other useful data sets) are available from the repository maintained by Naumann[6] [195]. Koepcke et al. [169] recently conducted an evaluation of several data matching systems using a set of four real data sets which the authors have made publicly available as part of their framework for evaluation data matching system, FEVER.[7]

It is important to note that these data sets provide only limited amount of information about the performance (with regard to matching quality and complexity) of data matching and deduplication systems or techniques. First, the data sets where the match status is known are all fairly small, therefore limiting the evaluation of scalability of a techniques or system. Second, each of these data sets contains a very specific type of data. Any results achieved on them should not be generalised to other types of data, even to data from the same domain but with different characteristics (such as different error characteristics or different attributes).

A comparison of different indexing techniques on three of the above listed data sets (Cora, Restaurant, and Census) is presented in Sect. 4.10, based on experiments recently presented by Christen [64]. As this evaluation illustrates, different indexing techniques perform quite differently on these three data sets with regard to the time required to build an index data structure, the number of candidate record pairs generated, and also the amount of memory required.

Despite all their limitations, the use of publicly available test data sets in data matching research has the advantage that researchers can (to some degree) compare their new algorithms and techniques to other existing algorithms and techniques. This is a much better approach for scientific progress compared to the use of proprietary or confidential data that cannot be given away, thereby making any evaluation of published research results and any comparison between techniques difficult. Ideally, research publications in data matching and deduplication that contain empirical evaluations should have been conducted on different data sets that are publicly available to illustrate how generalisable a novel data matching technique is.

7.6 Synthetic Test Data

An alternative to using publicly available test data sets, which have limitations in their size and content, is to generate data that can be used to test and evaluate data matching and deduplication systems or techniques. Such synthetic data should have characteristics that are representative for the real data on which a data matching system will be applied. This means that synthetic data should contain the same or at least similar attributes, the values in these attributes should follow frequency distributions close to those in corresponding real data, and the data should also have

---

[6] http://www.hpi.uni-potsdam.de/naumann/projekte/repeatability/datasets
[7] http://dbs.uni-leipzig.de/de/research/projects/object_matching/fever

similar error characteristics as one would expect in real data from the same domain. For example, an attribute that contains given names should contain strings that follow a frequency distribution similar to given names in a real database (such that 'Thomas' and 'Emily' occur more frequently than 'Aidyn' and 'Roberta', following current popular baby name distributions, as was discussed in Sect. 3.2) and also contain nicknames that can occur in given name attributes (like 'Bob' and 'Liz').

Generating such 'real' synthetic data can be a challenging undertaking. There are two basic approaches of how synthetic data can be created [56, 72].

- In the first approach, complete data sets are generated using (1) look-up tables that contain real attribute values and possibly their frequency distributions, and (2) rule-based techniques that generate attribute values according to rules that specify the length, distribution and content of these values. The first method is mainly suitable for attributes that contain a large number of different values, such as personal name and street and location name attributes. The second method, on the other hand, is suitable for attributes such as telephone, drivers or social security numbers which contain more structured values.

  Records consisting of a set of attributes can be generated using look-up tables and rule-sets appropriate to the content of the attributes. Various parameters have to be set by a user for example to specify the size of the database(s) to be generated and what look-up tables and rule-sets to use to generate the synthetic records. Because in real data there are commonly dependencies between attributes (for example, the given name of a person is highly dependent on their gender, while surnames depend upon the cultural background of an individual), it is of advantage if such dependencies can be modelled when data are generated. However, the more such dependencies are introduced the more complex the data generation process becomes, and the more parameter settings are required. The danger then is that a user simply leaves these parameters at their default values, rather than carefully adjusting them to their needs.

  In order to create variations of the generated records (which will then constitute the known true matching records or duplicates), variations and errors need to be introduced. Again, such errors need to follow the characteristics of real-world errors as much as possible. The conditions that govern such error imputation, such as manual keyboard data entry or optical character recognition, have been discussed in Chap. 3 in the context of data pre-processing. Essentially, the corruption of the generated records to create approximate matching records or duplicates needs to introduce variations that model the errors that occur in a real-world data entry process [72].

  Specific error parameters that a user should be able to set individually for each attribute should include: the likelihood for an attribute value to be modified in some way; the likelihood for an attribute value to be removed (i.e. set to a missing value); the likelihood to change an attribute value with a new value from the same attribute; and the likelihood to introduce character modifications such as edits (inserts, deletes, substitutions and transpositions), keyboard typing errors (replacing a character with a character neighbouring on a typical keyboard, such as 'z' and 'x'), or scanning errors (replacing a character with a similar looking character, like 'S' and '5'). For character level edits, the distribution of where modifications are applied should also be based on a parameterised model, because studies have shown that errors in real-world data do not occur randomly at any position. For manually typed names, for example, they occur mostly towards the end of names [214].

For certain types of modifications, such as nicknames and common name variations or misspellings, having large look-up tables of known variations of values can help to generate more realistic data compared to simply inserting random modifications.

- An alternative approach to generating synthetic data from scratch, based only on look-up tables and rule-sets, is to use a real-world data set that contains records with the required content. Such data can be sourced either from a database within an organisation (such as the name and address details of all customers from a cleaned data warehouse) or from a publicly available data source (such as voters registration lists or telephone directories which are available to the public in certain countries). Such real data sets are a realistic source of variations and frequency distributions of values.

However, because such data sets are generally well cleaned and deduplicated (one would hope especially for electoral rolls [9]), the same data corruption process described above needs to be employed to generate records that contain variations and errors that can be used in the data matching process.

Figure 7.6 shows three sets of example records that were generated with the FEBRL data set generator [56, 72]. This generator works by first creating a set of *original* records (indicated by the string 'org' in their record identifier) in the first step, followed by their modification into duplicate records (indicated by the string 'dup' and a number as there can be several duplicates generated from the same original record) in the second step. This generator allows a large number of parameters to be set so that data of different error characteristics can be generated [72].

Compared to using publicly available or proprietary and confidential data sets for testing and evaluating data matching or deduplication systems and techniques, the use of synthetic data has various advantages [72].

- Because the data have been explicitly generated, each record can be given a unique identifier and each modified record (approximate match or duplicate) that is based on a certain generated record can be given an identifier that refers back to the 'original' record it is based on. Therefore, when such data are used to test a data matching system, the match status of each candidate record pair is known and both matching quality and complexity, as discussed in Sects. 7.2 and 7.3, can be calculated. This allows the performance of data matching systems to be evaluated in detail.

- The size of the data generated and their characteristics with regard to content and variability (types and likelihoods of errors and modifications) can be fully controlled by the user. It is therefore possible to generate data that have very specific error characteristics, and to test and evaluate how well different data matching systems and techniques can handle such data. The scalability, a major challenge rec_id, age, given_name, surname, street_number, address_1, address_2, state, suburb, postcode

rec-1-org,   *33*, *madison*, solomon, *35*, tazewell *circuit*, trail view, *vic*, *beechboro*, *2761*
rec-1-dup-0, 33, madison, solomon, 35, tazewell <u>circ</u>,   trail view, <u>viv</u>, beechboro, 2761
rec-1-dup-1, 33, madison, solomon, 35, tazewell <u>crct</u>,   trail view, vic, <u>bechboro</u>, 2761
rec-1-dup-2,   , madison, solomon, <u>36</u>, tazewell circuit, trail view, vic, beechboro, <u>2716</u>
rec-1-dup-3, 33, <u>madisoi</u>, solomon, 35, tazewell circuit, trail view, vic, <u>beech boro</u>, 2761

rec-2-org,   *29*, soida, perera, *416*, marchant place, *weemilah*, *nsw*, belmont, 2280
rec-2-dup-0, 29, soida, perera, <u>414</u>, marchant place, <u>wemilah</u>, nsw, belmont, 2280
rec-2-dup-1, <u>92</u>, soida, perera, 416, marchant place, weemilah, <u>naw</u>, belmont, 2280

---

rec_id, age, given_name, surname, street_number, address_1, address_2, state, suburb, postcode

rec-3-org,   *29*, *jalisa*,   *wane*, 25, *prisk* place, *seabank*, , wa, latham, 6616
rec-3-dup-0, 29, <u>ghialisa</u>, wane, 25, prisk place, <u>zeabank</u>, , wa, latham, 6616
rec-3-dup-1, 29, <u>jalisa</u>   <u>whane</u>, 25, <u>prisc</u> place, seabank, , wa, latham, 6616
rec-3-dup-2, 29, <u>jalissa</u>,   wane, 25, <u>prisk</u> place, <u>seapank</u>, , wa, latham, 6616

rec-4-org,   39 , desirae, *contreras*, 44, maltby street, *phillip* lodge, nsw, *burrawang*, 3172
rec-4-dup-0, 39, desirae, <u>kontreras</u>, 44, maltby street, phillip lodge, nsw, <u>burrawank</u>, 3172
rec-4-dup-1, 39, desirae, contreras, 44, maltby street, <u>fillip</u> lodge,   nsw, <u>buahrawang</u>, 3172

---

rec_id, age, given_name, surname, street_number, address_1, address_2, state, suburb, postcode

rec-5-org,   28, *phyliss*,   winter, 20, *aspinall* road, , qld, *wairewa*, *3887*
rec-5-dup-0, 28, phyliss, winter, 20, aspinall road, , qld, wairewa, <u>3881</u>
rec-5-dup-1, 28, <u>phyl'lss</u>, winter, 20, <u>aspinall</u> road, , qld, <u>wajrewa</u>, 3887

rec-6-org,   *81*, *madisyn*, sergeant, 6, *howitt street*, creekside cottage, vic, *nangiloc*, 3494
rec-6-dup-0, <u>87</u>, madisyn, sergeant, 6, howitt street, creekside cottage, vic, <u>nanqiloc</u>, 3494
rec-6-dup-1, 81, <u>madisvn</u>, sergeant, 6, <u>hovitt</u> street, creekside cottage, vic, nangiloc, 3494

Fig. 7.6 Three examples of records created with the FEBRL data generator with different error types [72]. Original values that were modified are highlighted in bold-italics and their corresponding modified values are underlined. Two modifications were introduced into each duplicate record. The data used to generate these records consisted of name and address values taken from an Australian telephone database **a**. Typographic errors **b**. Phonetic errors **c**. OCR errors

in data matching and deduplication, can also be tested by generating data sets of different sizes.

- The generated data sets can be published openly so that other researchers can conduct comparative evaluations on these data sets and reproduce results from other research studies. This makes research in data matching and deduplication more meaningful compared to the situation where researchers only evaluate their new algorithms and techniques on their own (possibly not published) data sets.
- The program used to generate synthetic data can be published itself, allowing other researchers and practitioners working in the field of data matching and deduplication to generate their own data that are tailored specifically to their need. For

example, data specific to a country, culture or language can be easily generated by using appropriate look-up tables and parameter settings for errors and variations.

Even though synthetic data have all these advantages, the main problem with such data is still that, even with sophisticated look-up tables, attribute dependency and corruption models, such data will never be able to fully represent all the intrinsic characteristics of real-world data that make accurate and efficient data matching and deduplication such challenging problems.

Several data generators specifically aimed at generating data for data matching and deduplication have been developed. A first such generator was presented by Hernandez and Stolfo in 1995 [140]. It is known as UIS DBGen and is available from the RIDDLE repository.[8] This generator allows a user to create records and duplicates of these records using lists of names, cities, states and postcodes. It however cannot deal with frequency information for these values. This means that the frequency distribution of values will be uniform and therefore not follow the likely frequency distributions of real data. A user can set the number of records that are to be generated, the percentage and distribution of duplicates to be generated, as well as the types and amounts of errors to be introduced.

An improved generator was described by Bertolazzi et al. in 2003 [29]. It allows parameter settings that control if values in certain attributes become missing, and it also improves the variability of the created values by providing a larger number of modification and error types that can be introduced when duplicate records are generated. It is not clear if this data generator can handle frequency information, as not many details were published by the authors.

The FEBRL data matching system, described in detail in Sect. 10.2.4, includes a data generator [56, 72] that improves both the generator developed by Hernandez and Stolfo and the one developed by Bertolazzi et al. The FEBRL generator allows many parameters to be set with regard to the types, locations and likelihood of modifications applied to attribute values when duplicate records are created. Besides frequency look-up tables of attribute values, this generator also allows nickname and name variation look-up tables, as well as the specification of individual probabilities of certain types of errors, such as phonetic, keyboard and optical character recognition (OCR) errors, as Fig. 7.6 illustrates. The latest version of this generator [72] also allows dependencies between attributes to be specified (such as between a gender and a given name attribute), and it can even generate groups of records that correspond to a family. For such groups, the number of records in them, as well as their age and gender values, are drawn from specific distributions to allow realistic generation of parents and their children.

A generator similar to the one implemented in FEBRL was recently described by Talburt et al. [250]. This generator allows the creation of sequences of records that correspond to the occupancy of people as they live at different addresses over a certain period of time. The generator first creates a record that corresponds to an individual based on real data (such as publicly known real addresses). Two scenarios are possible, the first is for a single individual while the second scenario models couples

---

[8] Available from: http://www.cs.utexas.edu/users/ml/riddle/data.html

living together. Sequences of records (each with a time stamp) are then generated for each individual according to the selected scenario and by introducing variations into both the name and address attribute values over time based on variations collected from real data sources.

Other data generators that can create or corrupt data in XML format have also been developed [195]. Further to the generators used in data matching and deduplication, generators for specific types of data (such as relational database tables or biological sequences) have been developed by researchers in their respective communities.

7.7 Practical Considerations and Research Issues

The most important consideration when evaluating the outcomes of a data matching project is if ground-truth data (gold standard) in the form of known true matches and true non-matches are available or not. If no such training data are available, then the next question is if there is a practical way to obtain or create such data that are of high quality within a reasonable amount of efforts. In some circumstances, a data matching system can be assessed using either synthetically generated data or using one of the various test data sets that have been published. For both of these the true match status of record pairs is generally known.

As was described in Sect. 7.2, the commonly used quality measure of accuracy should not be used in the context of data matching, due to the much larger number of non-matches compared to matches that are normally contained in the set of candidate record pairs. Precision and recall, as well as the f-measure, are suitable measures to assess data matching quality.

Besides the actual run-time of a data matching system on certain data sets, the number of candidate record pairs generated, or the measures of reduction ratio, pairs quality, and pairs completeness, allow the measurement of the complexity of a data matching system and its effectiveness. These three measures also allow hardware-independent comparisons of different data matching systems.

When synthetic or publicly available test data sets are used to evaluate a data matching system or technique, then it is important to be aware of the limitations of such data, and the results achieved with them should not be generalised. The performance of a data matching system or technique is dependent on the type and the characteristics of the data that are matched. Having good domain knowledge will be of high value to achieve good matching or deduplication results.

Research efforts should be aimed at developing large test collections for data matching and deduplication in a similar fashion as has been accomplished in areas of information retrieval (such as the Text REtrieval Conference (TREC) data collection[9]), or machine learning and data mining (for example the University of California Irvine (UCI) repository[10]). Such data collections should contain both synthetic and

---

[9] http://trec.nist.gov/

[10] http://archive.ics.uci.edu/ml/

real-world data sets if feasible. Data containing personal information generally cannot be made publicly available due to privacy concerns as will be discussed in the following chapter.

An alternative to a data repository is the development of a test environment where researchers can upload their data matching algorithms. These algorithms are then evaluated on different data sets against a set of benchmark algorithms [195]. The results of such evaluations are being returned to researchers and potentially also published on a type of leader-board, indicating the performance achieved by different algorithms on various types of data matching problems.

In order to allow such a test framework to operate, an implementation-independent description of data matching algorithms is required. Similar to the predictive model markup language (PMML) initiative by the Data Mining Group,[11] an XML-based descriptive language of data matching algorithms would need to be developed.

7.8 Further Reading

The introductory book on duplicate detection by Naumann and Herschel [195] nicely covers the topics of quality measures, real-world and synthetic data sets, and data generators. The authors also discuss issues related to benchmarking data matching and deduplication techniques.

Christen and Goiser have provided a book chapter [71] that discusses in detail the issues involved in measuring data matching quality and complexity, and they provide an overview of a number of different quality measures. More recently, Menestrina et al. [187] discuss a novel approach on how to measure the quality of the record merging step, where records that have been classified as matches are merged into new combined records.

---

[11] http://www.dmg.org/

Part III
Further Topics

Chapter 8
Privacy Aspects of Data Matching

8.1 Privacy and Confidentiality Challenges for Data Matching

Existing data matching techniques assume that all data that are required for the matching are available in their original form (unencoded and unencrypted) to the organisation that conducts the matching. In many applications the databases to be matched contain records about individuals (such as patients, customers, travellers, students, etc.). Due to the lack of unique entity identifiers, the matching of such databases needs to be based on the personal details of the individuals whose records are stored in these databases. These details can consist of people's names, addresses, telephone, social security or driver's license numbers, or their dates of birth or death.

If data matching is conducted within a single organisation and between databases owned by this organisation, then privacy and confidentiality are generally not of concern. One can assume that the individuals who conduct data matching projects within an organisation are aware of all relevant policies and regulations with regard to handling the private and confidential data that are being matched, and that they do not have malicious intents to take identifying or other sensitive information, or the matched data, outside of their organisations for personal gain.

If matched data are used for purposes internal to an organisation only, such as for internal fraud detection, generating customer mailing lists, or internal research studies, then no privacy or confidentiality concerns will occur, assuming the necessary steps have been taken to prevent unauthorised access to the matched data and no detailed results of a matching exercise are made public.

On the other hand, if matched data are being passed on to another organisation, or if (parts of) the matched data are to be made publicly available, for example to researchers at different universities as part of a public health study, then the relevant privacy and confidentiality regulations need to be taken into account. In many countries, only data that do not allow the identification of individuals can be made publicly available [194]. In the USA, for example, health data that contain identifying information are covered by the Health Insurance Portability and Accountability

Act (HIPPA), which regulates what kind of matching and analysis can be conducted with health data, and at what level of detail health data can be published. Gliklich and Dreyer provide an extensive discussion of the various aspects that need to be considered when data from health registries are matched [122]. Laws similar to the HIPPA govern data publishing and sharing in many other countries.

Privacy concerns also arise when data matching is being conducted across databases that are held by different organisations, and when the matching requires identifying data to be shared and exchanged between organisations. As will be shown in Sect. 8.2, there are various scenarios where such inter-organisational data matching can lead to great benefits, but where privacy and confidentiality concerns or regulations limit or even prevent data matching projects.

The two major aspects with regard to privacy and confidentiality when databases are matched across organisations are that (1) traditional data matching techniques require that all data needed for the matching are given to the organisations that undertake the matching, and (2) the results of a data matching exercise using data from different organisations can reveal sensitive or confidential information that is not available in the data held by a single organisation [58].

8.1.1 Requiring Access to Identifying Information

If data that contain personal details need to be communicated from the organisation that holds the database needed for matching (named the *database owner*) to the organisation that will conduct the matching (named the *matching unit*), then the database owner will loose control over their data. Access to sensitive and confidential information and how it can be used is generally regulated through laws, and by policies and procedures internal to an organisation. Additionally, confidentiality agreements are commonly signed by all parties that are involved in a data matching project and by the individuals who have access to the data to be matched. Public research organisations furthermore have institutional review boards or ethics committees that assess projects that deal with personally identifying information. They only approve research where the public benefit outweighs the risk to privacy or confidentiality of the individuals whose records are stored in the databases to be matched.

While data are generally encrypted during the communication between organisations, if the matching requires access to the original unencrypted data then the database owner needs to trust the matching unit to be able to maintain the secure storage and processing of their data, and limit access to their data only to individuals involved in a data matching project. The security of the IT system at the matching unit must be able to prevent any unauthorised external access, as well as assure that no unauthorised internal access can occur.

Ideally, therefore, would be data matching techniques that do not require that any identifying data from a database owner need to be given to any other party involved in an inter-organisational data matching project, while still allowing that accurate data matching can be achieved on large databases. Such privacy-preserving data matching techniques will be discussed in Sect. 8.3.

8.1.2 Sensitive and Confidential Outcomes from Matched Data

The record pairs classified as matches in a data matching project can contain information that is not available in the individual source databases that were matched. The combined information can be highly sensitive. As an example, assume different health databases are matched and it is detected that a high profile individual (such as a politician or movie star) has a serious health problem in the form of a contagious disease [53]. While such highly sensitive information will not be made public directly, there is a risk that it is being revealed through gossip or by somebody taking personal advantage of this information. Another example occurs in situations where the matching is conducted with the aim to identify individuals who might have been involved in crimes or fraudulent activities, or who might be planning terrorist attacks. Here a falsely matched pair of records could result in an individual being accused of a crime they did not commit, or being added to a terrorism watch list. This can severely impact on somebody's life, including their travels or credit worthiness [153].

Even when identifying details are removed from matched data, for example before they are given to researchers for further analysis or made publicly available, it is possible in certain situation that individual entities can be re-identified [180]. This was demonstrated by Sweeney, who successfully matched "de-identified" medical records that were publicly available back to individuals, including the governor of Massachusetts [247, 248]. Sweeney also showed that even when using only the three attributes zipcode (5-digits), gender, and date of birth, nearly 90 % of the population of the USA (216 out of 248 million individuals) had a unique combination of these three values. Therefore, these three attributes can be used to uniquely identify nearly everybody in the population of the USA. Generally, even an attribute that is deemed not to be identifying can, in combination with other attributes, become identifying [194].

Not only can matched data be sensitive with regard to individual records and the entities they represent, but also with regard to groups of records that refer to groups of individuals with certain characteristics. For example, assume the matching of two databases, one containing medical details of patients and the other containing socio-economic information such as income, race, and the cultural background of people. The results from matching these two databases might reveal that there is a significantly higher prevalence of a certain contagious disease for people from a specific racial or cultural background living in a certain area. If such a result becomes public knowledge it could lead to stigmatisation of this group of people, and diminish their chances of obtaining employment or certain types of insurances.

How to prevent the re-identification of individuals from matched data that have been made public is an activity that has been investigated by research into statistical disclosure control [92, 253, 254]. By applying data matching techniques and using

different publicly available pieces of information, it can be shown that in certain
cases high re-identification rates are possible [92]. Such re-identification techniques
can also be used by criminals whose aim is to collect and match enough identifying
information to allow them to commit identity fraud [10, 206].

8.2 Data Matching Scenarios

While matching databases across different organisations can lead to benefits in application areas such as health, crime and fraud detection, and national security, as was
discussed in the previous section there are concerns about the sharing and matching
of databases that contain personal information across organisations. Matched data
has the potential to reveal sensitive private or confidential information that is not
available otherwise, information that can be misused and that can lead to damages
to both individuals and organisations [58, 75, 79, 134]. This can lead to the rejection
of data matching techniques, and result in the reluctance of organisations in both the
private and public sectors to employ data matching.

The following scenarios from a diverse range of application areas illustrate the
impact data matching can have on privacy and confidentiality when data are being
matched between different organisations [58]. Various real-world stories related to
privacy and data matching have also been described by Clifton et al. [79] and by
Fienberg [110].

- *Public health research*: A research project aims to investigate the effects that
  different types of car accidents have on the public health system. The questions
  this research aims to answer include which kinds of injuries are most common
  with what type of car accident; the characteristics of the drivers and cars involved
  in serious accidents; when and where accidents occurred, and what the road and
  weather conditions were at the time of the accident; the general health of people
  who were involved in a car accident one, two, five and ten years after their accident;
  and the financial burden of different types of car accidents upon the public health
  system. Such research can lead to policy changes that potentially can save many
  lives [44, 78].
  To enable this research, the research team requires data from hospitals (both public
  and private), doctors (general practitioners as well as specialists), private health
  insurers, government health departments, private car insurers, road and traffic
  authorities, and the police. Legal and commercial restrictions will prevent some
  of these organisations from sharing their data with the research team. A technique
  is therefore needed that allows matching all of these databases such that (1) no
  database owner is able to learn anything about any of the other databases that are
  being matched; (2) no external adversary is able to learn anything about the source
  databases even if they would get access to any data exchanged between the parties
  involved in the matching; and (3) only selected attributes of the matched records

are revealed to the research team (such as age and gender, and medical, car and accident details, but not names or addresses).

The current practice for dealing with such situations is that all required databases are sent to a trusted data matching unit, which, for example, is run by a government health agency [44]. This data matching unit will conduct the matching and send the required attributes of the matched records to the research team. All communications between the organisations involved in such a protocol will be encrypted, and all parties involved will have strict confidentiality requirements they need to follow, and the research will require approval by an institutional review board or ethics committee. Under such a strict regulatory framework the above described research project might be feasible. If a best practice approach is followed [161], only personal identifying information will be sent to the data matching unit to conduct the matching, but medical or otherwise confidential information of the matched records will be directly given to the research team [73].

Various countries have successfully set-up such institutional data matching units and corresponding matching protocols [161]. However, the weakness of such an approach is that all identifying data need to be sent to the organisation which will conduct the matching. While the communication is encrypted, the actual matching can only be conducted on the original unencrypted attribute values. The security of such a data matching unit therefore needs to be very high, and complete trust is required by all parties involved that the individuals who conduct the matching will not misuse the data they have access to, that proper access controls have been arranged, and that the IT systems used for the matching are secure. Any breach and compromise of sensitive information will likely lead to a backlash by the public which could result in the closure of such data matching activities in the area of public health.

- *Business collaborations*: Two companies plan to consolidate their businesses. They would like to know how many customers and suppliers they have in common, without having to share their complete confidential databases with each other. A technique is required that identifies the common entities stored in their databases, without revealing any other information to either business. Because of different database schemas and different formatting and encoding standards, some form of approximate matching is required to identify pairs of records from the databases of the two businesses that have high similarities. For confidentiality reasons and because of the danger of collusion, the involvement of a third party to conduct the matching of the data is not desired [79].

  In this scenario, techniques are needed that allow the sharing of large amounts of data between two organisations in such a way that (1) only records in the two databases that are similar with each other according to some similarity function and corresponding minimum similarity threshold are identified; and (2) the identities of these records and their similarities are revealed to both organisations. Neither of the two parties must be able to learn anything else about the other party's confidential data.

  Compared to the public health scenario discussed previously, the competitive commercial environment might prohibit the use of a third party to conduct the matching, because collusion between one of the database owners (the first business) and the matching unit (or an employee of the matching unit) could reveal all confidential information of the second database owner (the other business) to the first one. The potential of security breaches by intruders at the matching unit could also be deemed to be too high a risk. This scenario therefore requires a secure two-party data matching protocol.

- *National security*: For a terrorism related investigation, a national security agency requires access to both commercial and government databases, for example from financial institutions, car rental companies, airlines, immigration agencies, and residency agencies. If these databases would be queried by the national security agency with the details of suspect individuals, then their identities would be revealed to the owners of all databases that are being queried. This could compromise the investigation and seriously affect the lives of the individuals under investigation (of whom most will be innocent).

  A technique is required that allows querying these different databases such that neither the details of the query nor the answers (the matching records) are being revealed to the database owners [55]. The actual matching of individual records (that contain details about an entity) with several very large databases that contain records of many millions of entities needs to be conducted in real-time to provide fast query responses. Section 9.3 will further discuss the challenges of real-time data matching.

- *Geocode matching*: Cancer registries in many countries collect information about the occurrences of different types of cancer. Medical information about cancer patients is often collected together with their demographic details (their age, gender, address, and so on). Cancer registries are often small organisations that are partially funded by governments but otherwise rely upon public and private donations. Imagine such a cancer registry plans to match the addresses in their database to geographic locations (a process known as "geocoding," a topic that will be further covered in Sect. 9.1) [228]. The aim of such a geocoding project is to create a database that allows the spatial analysis of where different types of cancer occur, and if there are clusters of certain types of cancer or correlations with environmental factors (such as a chemical factory or nuclear facility that is located in proximity to where a group of cancer patients live or work). Because of their limited financial resources, the cancer registry cannot invest in an in-house geocoding system (i.e. software and personnel) but has to rely on an external company to conduct the geocode matching.

  The legal or regulatory framework might not allow the cancer registry to send their data to an external organisation for geocoding. Even if allowed, complete trust is needed in the capabilities of the geocoding company to conduct accurate matching, and to properly destroy the registry's address database afterwards. An alternative approach (if available) would be for the cancer registry to use the geocoding service offered by a trusted external organisation such as a government health department. In both cases, the addresses of cancer patients have to be made available to the outside organisation that performs the geocoding.

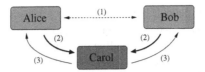

Fig. 8.1 Schematic view of a basic three-party protocol for privacy-preserving data matching. In the first communication step, the two database owners, Alice and Bob, exchange in a secure fashion information such as the parameter values and functions used in the protocol. In the second step they send their (somehow) encoded databases to the third party, Carol, which will conduct the matching. In the third communication step, Carol returns information about which records were classified as matches back to the two database owners. The exact content of each of these communication steps, and the computations done by the parties involved in the protocol, depend upon the actual privacy-preserving data matching protocol employed, as described in Sect. 8.3

This scenario requires a technique that facilitates accurate matching of addresses to locations in a secure way, such that no details of the addresses stored in the cancer registry's database are revealed to any external organisation [63]. The matching needs to be able to take variations, errors and missing values in addresses into account and allow some form of approximate matching suitable for address details.

As these scenarios highlight, there is a need for techniques that allow data matching in such ways that the database owners do not have to reveal any of their sensitive private or confidential data to any other organisation involved in a matching exercise, and only the matched records are being disclosed to the organisation(s) or individual(s) that require them. Research into such privacy-preserving data matching techniques has been conducted over the past decade, and the following section provides an overview of these techniques.

8.3 Privacy-Preserving Data Matching Techniques

The problem that privacy-preserving data matching aims to address is how two or more organisations can determine if their databases contain records that refer to the same real-world entities without revealing any information besides the matched records to each other or to any other organisation [104].

As the scenarios in the previous section have illustrated, there are two basic protocols of how privacy-preserving matching between two organisations can be achieved, as will be described below. Following the notation used in the cryptographic literature [237], the two owners of the databases that are to be matched are named *Alice* and *Bob*. Note that they do not need to correspond to individuals but more commonly they refer to organisations such as businesses or government agencies.

- *Three-party protocol*: In the first protocol, a third party, named *Carol*, is employed to conduct the matching. The general approach of this protocol is illustrated in Fig. 8.1. It consists of three primary communication steps (individual protocols might have additional steps). In the first step the two database owners exchange details such as which database attributes to use for a matching, the techniques to be used to pre-process and encode or encrypt their data, and any encoding or encryption keys required. These encoding or encryption keys must be kept secret from any other party and any external adversary or attacker. In the second step the (somehow) encoded databases are sent to the matching unit, Carol, which conducts the matching by calculating the similarities between records. In the third step the identifiers of the record pairs that have been classified as matches (i.e. the records that have high similarity with each other) are sent back to the two database owners, Alice and Bob. Depending upon the aim of the data matching exercise, the two database owners can now decide to either exchange details of the matched records, or to send the values of selected attributes of the matched records to another party, such as a research team.

The use of a third party is a general drawback of three-party protocols. Even though Carol does not receive any unencoded data (and therefore does not learn the content of individual records, such as their attribute values), there is the potential that one of the database owners (who knows the functions and keys used to encode the data before they are sent to Carol) is colluding with the matching unit. This would allow the colluding parties to decode all private and confidential information of the other database owner. Another way of how confidential information could be revealed is when one of the database owners is able to get access to Carol's IT system in a malicious way, which will again allow this database owner access to the other database owner's confidential information. A third way of a security breach occurs if an external attacker gets access to the IT system of one database owner and also to Carol's IT system, which will allow this attacker to infer information about the second database owner's data.

- *Two-party protocol*: An alternative to three-party protocols is to remove the third party such that only the two database owners communicate directly with each other, as illustrated in Fig. 8.2. Two-party protocols also have a first communication step where details are exchanged about the attributes and the pre-processing and encoding or encryption functions that will be used in the protocol. The second and third communication steps are similar to the steps in three-party protocols, however in two-party protocols these steps are often repeated several times. While two-party protocols are generally more secure (because there is no possibility of collusion with a third party), they often require more sophisticated encoding or encryption techniques because the two database owners must hide their private or confidential data from each other. This makes two-party protocols computationally more complex in general.

Various techniques to facilitate privacy-preserving data matching have been investigated. Some of the key concepts used include one-way hash-encoding, secure multi-party computation (SMC) approaches such as commutative and homomorphic encryption and split data, Bloom filters, differential privacy, mapping attribute values into multi-dimensional spaces, and the use of public reference values. These concepts will be explained and illustrated further in the following sections.

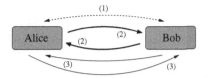

Fig. 8.2 Schematic view of a basic two-party protocol for privacy-preserving data matching. The protocol consists of the same three basic communication steps as the three-party protocol shown in Fig. 8.1. Step two is often repeated several times, as many two-party protocols are based on only exchanging parts of their data in each communication step. The actual data exchanged and the way they are encoded or encrypted depend upon the specific privacy-preserving data matching protocol employed, as described in Sect. 8.3

There are two security models of how the parties in a protocol are assumed to behave. In the first model, known as the "semi-honest" [123] or "honest but curious" [134] model, it is assumed that all parties that take part in a protocol follow the steps of the protocol, but they store all the data they receive from any other party and all the results of the computations they carry out. Using this information, a party then might try to infer any sensitive knowledge of any of the other parties. For example, the data communicated might allow a party to compile statistics of the frequency distributions of (encoded) attribute values, which can be used in a frequency analysis (also known as a frequency attack) where the frequencies of external data (for example name values from a public telephone directory) are matched with the frequencies of the encoded attribute values.

In the second security model, known as the "malicious" model [123, 134], it is assumed that the parties that take part in a protocol do not necessarily follow the steps of the protocol, but they might maliciously communicate either wrong or modified data in order to be able to infer any sensitive information from any of the other parties. For example, assume prior to starting the matching protocol, database owner Alice inserts an extra record with specific values (such as a name "John Dijkstra") into her database (even though no individual with these values exists in her database). If this record is then matched with a record from Bob's database, then Alice has learned that Bob's database contains a record with these specific values (i.e. there is a record for "John Dijkstra" in Bob's database).

Most approaches to privacy-preserving data matching have been developed assuming the semi-honest rather than the malicious model, because the former holds for many real-world situations where all parties involved in a data matching exercise have an interest in achieving the outcomes of the matching rather than maliciously trying to deceive the other parties. Additionally, protocols developed for the malicious model generally have much higher communication and computation complexities than protocols developed for the semi-honest model, because the model requires much more sophisticated encryption and encoding technology to assure any malicious behaviour is detected or prevented [123].

String	SHA1 hash	MD5 hash
Peter	64ca93f83bb29b51d8cbd6f3e6a8daff2e08d3ec	6fa95b1427af77b3d769ae9cb853382f
peter	4b8373d016f277527198385ba72fda0feb5da015	51dc30ddc473d43a6011e9ebba6ca770
Pete	ec81523bfe26c5232d53a01df4efae048badf1d1	fbf3589622d820bd28023c7b2d7b91ac
pete	e3b6cda228242c30b711ac17cb264f1dbadfd0b6	858d41c9e397b8fa34bb046d8055f276
gail	18a6b6530e7d44721f6f34abe94a0522fd4265c5	9d1568a5ff0b38c3b26b52ed987c36db
gayle	dde891995ff68fdfd84b08dea441e9df3f63b6e6	1c73180795ad6ee7dc6a6d342b23dbd2
gale	b449ee4c45fcb7757e9582ced52f803f22153791	fdbd3cd60f63ebe9505bb7e0310a73d2
gael	e7582b9507331a5564b63863c9f53d17cb7fc228	87931780ab7fb3123b2d1ce18a95970e

Fig. 8.3 Example hash-codes using the SHA1 and MD5 one-way hash-encoding functions [237]. As can be seen, even a single character difference, or a different ASCII code (upper/lower case letter) results in a completely different hash-code. This makes the approximate matching of strings using their hash-codes unfeasible

8.3.1 Exact Privacy-Preserving Matching Techniques

The idea of conducting data matching between different organisations without the need of any identifying data having to be exchanged between the organisations was first explored by French health researchers in the mid 1990s [219, 220, 221]. One proposed approach is to convert each input string into a code based on polynomial coefficients such that different strings are represented by different codes [221]. With this approach, it is important that the likelihood of two strings being converted into the same code (known as a "collision" in the context of hashing) is minimised, and also that an adversary cannot infer the input string from the code only.

This initial approach was then improved by using a one-way hash-encoding function, such as the "secure hash algorithm" (SHA) or a "message digest" (MD) function [237], for the encoding of the input strings [219, 220]. A one-way hash-encoding function allows the efficient encoding of an input string into a hash-code, however having only access to a hash-code it is impossible with current computing techniques to find the plain-text string in a reasonable amount of time.

Figure 8.3 shows the SHA1 and MD5 hash encodings for two groups of example strings. As can be seen, even a single character difference (in the ASCII code value of a character) results in a completely different hash-code. The straight application of one-way hash-encoding therefore only allows exact matching. A possible way to improve this is to pre-process and standardise the input strings in a similar way as is done for phonetic encoding algorithms, as was discussed in Sect. 4.3. The idea of such pre-processing is to convert the input strings into new strings that are the same for similar (sounding) names. Name variations are converted into a standardised form using look-up tables (for example, "gael," "gayle" and "gale" can all be replaced by "gail"), and nicknames need to be expanded into their full names (so "bob" becomes "robert" and "liz" becomes "elizbeth"). Quantin et al. [221] present a pre-processing approach specialised for French names that consists of eleven steps. While such pre-processing can improve the matching quality for data that are dirty, it can also result in a higher false match rate (see Sect. 7.2), as more record pairs are classified as matches because their original attribute values have been pre-processed into the same standard form.

Once the hash-encoding has been applied by both database owners on the attributes that are used for the matching, only the hash-codes together with their (encrypted) record identifiers need to be communicated, by sending them to the matching party, Carol, in a three-party protocol. The matching of two databases is then based only on hash-codes that are exact matches. If several attributes are used in the matching, then for each attribute where the hash-codes are the same for a given record pair, the corresponding similarity for this pair is increased (by 1 if normalised similarity values are assumed). The pairs with a total summed similarity above a certain threshold are then classified as matches, in the same way as was discussed in Sect. 6.2.

Within a two-party protocol, the hash-encoding of attribute values does not hide information from the two database owners about the values in their respective databases. Because both database owners know the one-way hash-encoding function used, they can mount a "dictionary attack" [237] on the hash-codes received from the other database owner. A dictionary attack works by a party hash-encoding all the values in its database, or potentially even the values obtained from an external source (such as a public telephone directory that contains most given name and surname values in a country). The party can then conduct exact matching of these hash-codes with all hash-codes it has received from the other party, and thus it learns which values occur in records of the other party. Because of this, simple hash-encoding approaches to privacy-preserving matching can only work for three-party protocols.

However, in a three-party protocol even the matching unit can potentially mount a dictionary attack using external data sources by applying known hash-encoding functions (such as SHA1 or MD5) on a public database (such as a telephone directory) and checking for matching hash codes. A way to overcome this problem is to add a secret key k to each string s in a database before it is hash-encoded. This key needs to be agreed upon by the two database owners in a secret way at the beginning of the protocol such that the matching party, Carol, does not know the value of the key k. Rather than hash-encoding a string s by itself, $h = H(s)$, with $H(\cdot)$ being the one-way hash-encoding function (like SHA or MD5) and h the resulting hash-code of s, the string concatenated with the secret key is encoded, $h = H(k \oplus s)$, with $\oplus$ denoting a string concatenation. Because the matching unit, Carol, does not know the value of k she cannot mount a dictionary attack.

However, the matching unit can still mount a frequency attack, where she counts how often a certain hash-code h occurs in a certain attribute in the data that she receives from the database owners. She can then compare the distribution of these frequency counts with the frequency distributions of attribute values in a public database. If for example the frequency distribution of the hash-encoded values in an attribute are similar to the frequency distribution of surname values, then she can identify that the most commonly occurring hash-codes correspond to the most commonly occurring surname values (which might be "Smith" or "Miller"). Based on this information, the matching unit can aim to identify the value of the key k by trying a large number of possible values for k until she finds a k that together with a given hash-encoding function leads to hash-codes that match the most frequent hash-codes from the database owners.

Agrawal et al. [6] view the problem of matching values across two databases as the intersection of two sets of attribute values. They developed several two-party protocols that allow this set intersection to be computed securely, such that only the values in the intersection become available to the two database owners. Their protocols are based on a commutative encryption scheme, where for two encryption functions, $f(\cdot)$ and $g(\cdot)$ it holds $f(g(s)) = g(f(s))$ for any value of s. Using such an approach, neither of the two parties can calculate the encryption of a value s without the help of the other party. Both set intersection and equijoin protocols are then presented, and their security and their communication and computation complexities are analysed [6].

Privacy-preserving matching in the context of secure searching of keywords in databases was explored by Song et al. [244]. Their approach was to encrypt each word (assumed to be of a fixed length of characters) using a stream cipher (i.e. a pseudo-random number generator) [237] such that Bob, who will store the encrypted database, does not learn anything about the actual words in that database, and he also does not learn for which word Alice is searching when she is querying the database. This scenario is desirable when sensitive data need to be stored on a remote server or in the cloud. While the basic approach developed by Song et al. only supports exact querying of words, they discuss the possibility of wildcard searches of regular expressions based on the idea of generating all possible instantiations of a regular expression, and querying the database with all of them. For example, assuming only lowercase letters are used, a regular expression of the form "?bc," where the "?" represents a single character, would be expanded into the 26 query words "abc," "bbc," "cbc," ..., "zbc." This approach can clearly only be applied on a very limited set of wildcards, and it does not allow full approximate matching. This approach of generating all possible forms of a query can also be applied to an edit distance based approach. For example, again assuming only lowercase letters, with a three characters long string and a maximum edit distance of 1, a total of 182 variations of this string will need to be generated (3 deletes, 25 substitutions per character, and four positions to insert a character). For a four characters long string, the number increases to 234 variations, and with five characters there will be 286 variations. From the number of queries Alice will post for a wildcard search, Bob will be able to learn some information about the original string.

A different line of work was conducted by O'Keefe et al. [202], who developed protocols that allow the secure extraction of records from database tables such that the database owner does not learn which records were matched and extracted. The protocol requires three parties and is based on an "oblivious transfer" protocol [237] where the sender of a set of messages (records in the case of data matching) does not know which messages were chosen by the receiver of the messages, and the receiver obtains from the sender only the messages it selected but no others. While this protocol can solve a challenging problem for example in the heath area (to extract patients with certain sensitive characteristics from a database), it can only find exact matching values.

To summarise, while simple one-way hash encoding allows efficient privacy-preserving data matching across organisations, the main limitation of these approaches is that they can only find exact matches. Even if some form of data pre-processing is applied, only certain types of variations will be converted into the same hash-codes. As was however discussed in Chap. 3, most real-world data are dirty and contain a variety of errors and variations. To achieve high quality data matching within a privacy-preserving framework, approximate comparison functions, such as the ones presented in Chap. 5, need to be available for privacy-preserving data matching. Such techniques will be presented in the following section.

8.3.2 Approximate Privacy-Preserving Matching Techniques

Most research in privacy-preserving data matching so far has concentrated on developing techniques that allow the approximate matching of strings or sequences between organisations without the need that these values are being revealed to any party involved in the matching. Several surveys have recently provided comparative evaluations of the developed techniques [99, 158, 255, 262]. In this section a selected few key techniques are presented in order for the reader to gain an understanding of the main approaches to approximate privacy-preserving data matching.

In 2003, Attalah et al. [14] proposed a two-party protocol for securely calculating the edit distance (as presented in Sect. 5.3) between strings or sequences. The approach is based on a dynamic programming algorithm that generates a matrix that contains the number of edits required between two sub-strings, as illustrated in Fig. 5.1 on page 104. This matrix M is stored in a shared fashion between the two database owners, where Alice stores a matrix M_A (known only to Alice) and Bob stores a matrix M_B (known only to Bob) such that $M = M_A + M_B$. These two matrices are generated iteratively throughout the protocol, and at the end of the protocol only the final edit distance (the value in the lower right corner of the matrix M) is exchanged between the two parties. They therefore both learn the final edit distance between two strings but not the intermediate distances between sub-strings which would allow them to reconstruct the other party's string. Generating the matrix M is based on a homomorphic encryption approach and a minimum finding protocol for shared data. As a basic building block, for a vector of values $\mathbf{c} = \mathbf{a} + \mathbf{b}$, where Alice stores $\mathbf{a} = [a_1, a_2, \ldots, a_n]$ and Bob stores $\mathbf{b} = [b_1, b_2, \ldots, b_n]$, this protocol securely identifies which element in $\mathbf{c}$ is the minimum value without revealing any other information. In order to find out if $c_i \geq c_j$ is true or not, Alice and Bob can compare $a_i - a_j$ (known only to Alice) and $b_j - b_i$ (known only to Bob) which follows from $c_i \geq c_j \Leftrightarrow (a_i - a_j) \geq (b_j - b_i)$. This simple approach however reveals the index of the minimum value in a vector to the other party. This drawback can be overcome by applying a random permutation to the vectors before the comparison [14], which however increases the communication and computation complexities of the protocol. Using this basic secure comparison step, the calculations in Equation 5.5 on page 104 which find the minimum of three cells in the matrix M can be carried out such that neither Alice nor Bob learn each other's values. While this protocol allows the calculation of edit distance between two parties without the need to reveal the actual strings or sequences that are compared, the main drawback of the protocol is that it requires one communication step for each element of the edit distance matrix, and therefore a quadratic number of communication steps for the comparison of two strings or sequences.

In 2004, Churches and Christen [74, 75] presented a privacy-preserving approach to calculate the Dice coefficient (see Equation 5.10 on page 107) in a three-party setting. Named "blindfolded record linkage" [75], their approach is based on the idea of hash-encoding q-gram lists rather than full string values (as was discussed in the previous section), and to send these hash-encoded q-gram lists together with information about the number of q-grams in the strings to the matching unit (the third party Carol) which can calculate the Dice coefficient by counting the number of hash-encoded q-grams that two strings have in common (and multiplying this number by two) divided by the number of q-grams contained in the two strings. In order to allow fuzzy matching, not just complete hash-encoded q-gram lists are sent to the matching unit, but also sub-lists of these hash-encoded q-gram lists where one or more q-grams have been removed. For example, assume Alice has the value "peter" in her record "a1" and Bob has a value "pete" in his record "b2." Using bigrams ($q = 2$), Alice generates the bigram list ["pe," "et," "te," "er"] and Bob generates the list ["pe," "et," "te"]. Alice next generates all sub-lists of her bigram list with one bigram removed, and Bob does the same with his bigram list, resulting in the following bigram sublists (with the original bigram lists included):

Alice: ["pe," "et," "te," "er"], ["et," "te," "er"], ["pe," "te," "er"], ["pe," "et," "er"],
           ["pe," "et," "te"]

Bob: **["pe," "et," "te"]**, ["et," "te"], ["pe," "te"], ["pe," "et"]

The bigram sub-list in common is highlighted in bold font. Alice and Bob now each converts their q-gram sub-lists into hash-codes by first converting each list back into a string (and concatenating the string with a secret key) and applying a one-way hash-function that was agreed upon by the two database owners. They then send these hash-codes together with the number of q-grams in the sub-list, the record identifier, and the total number of q-grams in the string value to the matching unit. For the above example, and assuming simplified shortened hash-codes, Alice and Bob will send the following tuples to the matching unit Carol:

Alice: ("7d44721f6f34," 4,"a1," 4)
       ("23b2dce7d424," 3,"a1," 4)
       ("49ee4c45fcb7," 3,"a1," 4)
       ("6d342b23dbd2," 3,"a1," 4)
       (**"1c73180795ad"**, 3, "a1," 4)

Bob: (**"1c73180795ad"**, 3, "b2," 3)
     ("820bd28023c7," 2, "b2," 3)
     ("6fa95b1427a3," 2, "b2," 3)
     ("34bb046d8055," 2, "b2," 3)

For each hash-code that appears in both Alice's and Bob's tuples, the matching unit can now calculate the Dice coefficient between the string values these hash-codes are based on. For the given example, only the encoding "1c73180795ad" (shown in bold font), which is based on a three q-gram sub-list, appears in common. The Dice coefficient of these two tuples can be calculated as $\frac{2 \cdot 3}{3+4} = \frac{6}{7}$, which corresponds to the similarity between "peter" and "pete." 3 and 4 are the lengths (number of q-grams in the bigram lists) taken from the corresponding two tuples shown above. The biggest drawback of this approach to privacy-preserving approximate string matching is the need to generate sub-lists of q-gram lists. If the string values to be matched are long, then a large number of such sub-lists need to be generated. This results in a large number of tuples that have to be communicated to the matching unit. Experiments using an Australian database that contained surnames and suburb (town) names showed that the overhead of this approach compared to communicating the unencoded values only (for a non privacy-preserving matching) ranged from around 400 to over 4,300 [74].

Also in 2004, a two-party protocol for computing string distance measures such as SoftTFIDF (described in Sect. 5.8) as well as Euclidean distance was presented by Ravikumar et al. [226]. Their approach is based on a secure protocol for calculating the scalar or dot product between two vectors ($\mathbf{a} \cdot \mathbf{b} = \sum_{i=1}^{n} a_i \cdot b_i$). A secure scalar product can be calculated using a secure set intersection protocol [4] and a stochastic approach that samples elements of the two vectors whose similarity is to be calculated. Specifically, if $k < n$ is the number of samples to be used, then both database owners first normalise their vectors $\mathbf{a}$ and $\mathbf{b}$, respectively, and then sample k elements in these vectors such that the probability of an element being sampled is proportional to the value of the element. Each database owner then adds the indices of the sampled vector elements ($i \in \{1, \ldots, n\}$) into a set, and the secure set intersection protocol is applied on these two sets of indices. The resulting intersection of the two sets is multiplied by the normalisation coefficients to get the final result which approximates the dot product between the two vectors. Experiments on matching bibliographic records taken from the Cora data set (see Sect. 7.5) showed that with around 1,000 samples this secure scalar product achieves an accuracy comparable to the scalar product using the full vectors. The number of communication steps in this protocol scales linearly with the number of samples used.

Pang et al. [207] in 2009 proposed a three-party protocol that is based on the idea of using a set of reference values that are available to both database owners (for example given names and surnames extracted from a public telephone database). The two database owners individually calculate the distances between all of their attribute values and these reference values. They then send these distances together with the encoded attribute and reference values to a third party. For each unique pair of attribute values the triangular inequality (as discussed on page 102) is used to find an upper bound on the distance between the two attribute values. If more than one reference value is used then the minimum of the distance estimates is taken. For example, assume s_A and s_B are the strings held by the database owners, Alice and Bob, respectively. They both have access to two reference values r_1 and r_2. Independently they can calculate the distances (such as edit distance, discussed in

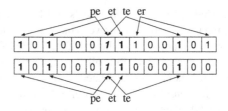

Fig. 8.4 An example pair of Bloom filters consisting of 14 bits and with each bigram being hashed with two hash-functions. As can be seen, a collision happens in the seventh bit (shown in italic font) where two different bigrams are hashed into the same bit. Five bits set to 1 (shown in bold font) are in common in both Bloom filters, with the Bloom filter that corresponds to the first string value ("peter") having seven bits set and the Bloom filter of the second string value ("pete") having five bits set. The Dice coefficient is therefore $\frac{2 \cdot 5}{7+5} = \frac{10}{12}$

Sect. 5.3 and denoted here with ed) between their string and the two reference values. Alice calculates $d_{A,1} = ed(s_A, r_1)$ and $d_{A,2} = ed(s_A, r_2)$ and Bob calculates $d_{B,1} = ed(s_B, r_1)$ and $d_{B,2} = ed(s_B, r_2)$. They then send the encoded attribute values (denoted by $e(s)$) and the distances to the matching unit, Carol, which can calculate an estimated distance $ed(e(s_A), e(s_B)) \le d_{min} = \min(d_{A,1} + d_{B,1}, d_{A,2} + d_{B,2})$. If the distance d_{min} lies below a given threshold then the two corresponding attribute values are classified as a match. The performance of this three-party protocol with regard to matching accuracy depends upon the set of reference values selected.

More recently, several researchers have investigated the use of Bloom filters [39] to conduct approximate matching within a privacy-preserving framework [98, 173, 239]. Bloom filters are bit-arrays, where a bit is set to 1 if a hash-function maps an element of a set (such as the set of all possible q-grams) into this bit. Formally, let l be the number of bits in the Bloom filter, and k be the number of independent hash-functions, $h_1(\cdot)$ to $h_k(\cdot)$, that are used to map elements of the set U into the Bloom filter. An element $u \in U$ (such as a q-gram) is hashed k times using $g_i(u) = h_i(u) \bmod l$, $i = 1, \ldots, k$, with $0 \le g_i(u) < l$ being the bit number in the Bloom filter that is set to 1. If a bit was already set to 1 it is left unchanged if it is set by another hash function (this is known as a collision). Figure 8.4 illustrates this approach, where a Bloom filter of length $l = 14$ bits and two hash-functions ($k = 2$) are used. Q-gram based similarities, as presented in Sect. 5.4, can be calculated based on the number of bits set to 1 at the same positions in two Bloom filters.

Schnell et al. [239] presented a three-party protocol using such a Bloom filter approach, and they showed that precision and recall results can be achieved using encrypted Bloom filters that are close to the results that can be achieved with unencoded q-grams. Durham et al. [98] confirmed that Bloom filter based matching can significantly outperform a binary comparisons approach. More recently, Kuzu et al. [173] showed that, depending upon the length of the Bloom filter and the number of hash-functions used, the Bloom filter approach proposed by Schnell et al. [239] can be attacked using a vigorous frequency analysis combined with a constrained satisfaction solver that allows the assignment of individual encoded values to unencoded attribute values from the two source databases. Durham in her recent thesis [100] examined how record-level Bloom filters, that are based on bits sampled from field-level (attribute-level) Bloom filters, can improve the security of privacy-preserving data matching.

8.3.3 Scalable Privacy-Preserving Matching Techniques

The techniques presented in the previous section provide solutions for the approximate matching of strings or sequences between organisations without the need of having to reveal these strings or sequences. These techniques however do not take into consideration the overall computation and communication complexity of the matching of databases that potentially can be very large. As was discussed in Chap. 4, non privacy-preserving data matching generally applies some form of indexing to reduce the number of record pairs that need to be compared. Recent research in privacy-preserving data matching has investigated how to improve the efficiency of privacy-preserving matching and make it scalable to large databases.

Al-Lawati et al. [8] in 2005 were the first to investigate indexing in the context of privacy-preserving data matching. They proposed three secure approaches to standard blocking (as presented in Sect. 4.4) within a three-party protocol. In this protocol, the similarities between records are calculated using the TF-IDF distance function based on hash-signatures. The blocking is based on tokens, for example q-grams, such that only record pairs that have a token in common become candidate record pairs that are being compared. In the first of the three proposed blocking approaches, called simple blocking, the data sets are blocked using the hash-signatures generated from the record values, and each record is inserted into several blocks, leading to a potentially large number of candidate record comparisons. The second approach overcomes this drawback by adding the record identifiers into the calculation of hash-signatures. This maintains the uniqueness of each record pair. In the third secure blocking approach, Alice and Bob first conduct a secure set intersection protocol to identify which blocks they have in common. They then only send these common blocks to the matching unit Carol, thereby reducing the amount of communication and computation required. Experimental results by the authors showed that this secure blocking can reduce the total run time required by up-to two thirds [8].

Scannapieco et al. [234] addressed both the problem of private schema and private data matching. The basic idea of their three-party approach is to map records into a multi-dimensional metric space using the SparseMap algorithm, which is similar to the approach on mapping based indexing that was discussed in Sect. 4.9. Strings are embedded into the metric space using a set of random reference strings. These reference strings are shared between the two database owners. The distances between the actual attribute values (assumed to be strings) and the reference strings in the metric space are then sent to the matching unit, which can calculate the similarities between values in the metric space. Matches are classified as the pairs of values that have a distance below a certain maximum threshold. The schema matching approach

Fig. 8.5 Example of k-anonymity generalisation of a small database table as used by the hybrid approach to privacy-preserving data matching proposed by Inan et al. [146] (figure adapted from [146])

RecID	Education	Age
r1	9th	22
r2	10th	16
r3	12th	27
r4	Masters	33
r5	Masters	39
r6	Masters	34

RecID	Education	Age
r1'	Secondary	[1–32]
r2'	Secondary	[1–32]
r3'	Secondary	[1–32]
r4'	Masters	[33–39]
r5'	Masters	[33–39]
r6'	Masters	[33–39]

works by the third party sending a global schema as well as a hash-encoding function to the two database owners, which can then map their own schema onto this global schema and hash-encode their local mappings. By using a secret key known only to the two database owners, each of them encodes their local schema mappings and sends them to the matching unit, which can calculate the intersection of attributes the two database owners have in common without knowing the names of these attributes.

Yakout et al. [290] converted the three-party protocol proposed by Scannapieco et al. [234] into a two-party protocol. In a first step, record pairs that likely refer to matches are generated by converting the values in the multi-dimensional metric space into complex numbers. These complex numbers are then exchanged between the database owners, so they each can generate the pairs of complex numbers that are within a maximum distance from each other. These are the record pairs that likely correspond to matches. In the second step of the approach, the two database owners calculate the actual distances between the vectors of all likely matched pairs using a secure scalar product protocol.

A different approach to scalable privacy-preserving data matching was presented by Inan et al. [146] in 2008 who combined data sanitisation techniques (such as k-anonymity or adding random noise) with a SMC protocol [123]. In this proposed three-party protocol, in the first step k-anonymity [248] is applied by the two database owners on their attributes that are used for the matching, resulting in modified databases where values have been generalised. Figure 8.5 shows an example of both an original and resulting generalised small database table. The hash-encoded generalised values are sent to the matching unit which can classify pairs into non-matches and potential matches depending upon how many generalised attribute values two records have in common. Non-matches are those pairs where the number of different hash-encoded values is above a certain threshold. For the potential matches, a more expensive SMC step based on homomorphic or commutative encryption is employed to calculate the actual similarity between record pairs [146]. An advantage of this approach is that a threshold setting allows a user to trade-off between precision and recall of the resulting matched record pairs.

More recently Inan et al. [147] proposed an approach to privacy-preserving data matching based on differential privacy. The aim of differential privacy is to provide accurate statistical information from databases while at the same time minimising the chances that individual records can be identified [101]. In the proposed two-party approach, differential privacy is employed in the blocking step (compared to k-anonymity used in the previous work by the same authors [146]), while the actual matching of records within each block is performed using SMC based on homomorphic encryption to calculate the Euclidean distance between attribute values. The approach uses specialised tree data structures to improve scalability. It however has a trade-off between accuracy and privacy [147].

Karakasidis and Verykios [157] have recently investigated how phonetic encoding algorithms, such as Soundex presented in Sect. 4.3, can be employed to accomplish scalable privacy-preserving approximate matching. While originally employed mostly in the indexing step to split the databases to be matched into smaller blocks, the one-to-many properties of phonetic encoding algorithms, their abilities to group similar values into the same phonetic code, and especially their computational efficiency make them suitable for privacy-preserving approximate matching. In this proposed three-party protocol, Soundex encodings are generated for several record attribute values. Each Soundex code is then replaced with a unique random number. To further improve the security, each database owner inserts extra faked records into their database. The encoded databases are then sent to the third party which can calculate the similarities between record pairs based on how many Soundex codes (represented by random numbers) a pair has in common. The identifiers of the record pairs that have a similarity above a certain threshold are then returned back to the database owners. In a follow-up paper, different strategies for adding extra faked records with different distributions of values is explored [159], highlighting that different data distributions can result in different levels of privacy preservation.

A two-party protocol that is different from all so far presented techniques was recently proposed by Vatsalan et al. [260]. Their idea assumes that the databases to be matched are large and therefore contain a large portion of all the possible values that can occur in an attribute in a population. For example, a large hospital database might contain most possible given names and surnames. The two database owners can therefore calculate the similarities between attribute values individually without having to communicate these values with each other or with a third party. Each database owner generates a data structure which contains these pre-calculated similarities. Using a set of reference values (similar as was proposed earlier by Scannapieco et al. [234] and Pang et al. [207]) and the triangular inequality and reverse triangular inequality then allows the database owners to find upper and lower bounds for the similarities between records. To hide the actual similarity values (which could reveal details about the actual attribute values), binned similarity values are exchanged between the database owners to identify which record pairs have a high similarity with each other. This approach was evaluated on a database containing nearly two million records, indicating its suitability for the privacy-preserving matching of large databases [260].

8.4 Practical Considerations and Research Issues

Privacy-preserving data matching is a relatively young area of research, and most techniques proposed so far (as discussed in the previous section) have not been implemented into commercial or other operational systems. There has been one commercial system by a big vendor that provides a technique for anonymous data matching, however not many technical details are available (some form of hash-encoding is applied in this system combined with sophisticated and domain specific data pre-processing). Practitioners who aim to implement privacy-preserving data matching are therefore likely required to implement a system by themselves. The three major criteria that need to be considered are (1) the sizes of the databases that are to be matched, (2) if a third-party can be employed to conduct the matching or not, and (3) how secure a protocol needs to be with regard to how much information about the databases can be revealed to the parties involved in the protocol.

There are still several challenges that have to be solved to make privacy-preserving data matching feasible for practical applications. The main areas that need to be investigated are scalability to matching large databases, being able to match different types of data, accurate and automated matching, and how to assess the completeness and quality of matched data in a privacy-preserving framework.

Being able to match databases that contain many millions of records in a privacy-preserving fashion is one of the most crucial issues that must be solved, because many real-world databases do contain such large numbers of records, and therefore it must be possible to match such databases in a reasonable amount of time. Most current techniques for privacy-preserving data matching have only been evaluated on small databases that contain less than a million records.

As was described in Chap. 5, data matching commonly relies upon the comparison of different types of data, including strings (such as names and addresses), numbers (like dates, ages or salaries), and other more specific data types (such as postcodes, or telephone, social security, or driver's license numbers). For privacy-preserving data matching to become practical, it must be possible that these different types of data can be compared in a secure fashion.

Techniques are required that allow the accurate classification of the compared record pairs into matches and non-matches within a privacy-preserving framework. Because it is unlikely that training data in the form of matches and non-matches are available, no supervised classification techniques can be employed. Privacy and confidentiality concerns will prevent that training data such as unencoded attribute values (like personal names and addresses) can be given to the party (or parties) that undertake(s) the matching. Rather, unsupervised techniques, such as clustering, need to be employed.

Finally, probably the most difficult challenge is how to assess the quality and completeness of the records that were classified as being matches within a privacy-preserving setting, where it is unlikely that the actual unencoded records will be available for manual inspection. Statistical estimates based on data distributions can provide some approximate value of matching quality. Another possible way to overcome this challenge would be to first conduct a matching using synthetic data that are closely modelled on the real data that are to be matched. However, as was discussed in the previous chapter, generating realistic synthetic data is in itself a formidable challenge. There has been one publication recently by Barone et al. [17] who investigated how accuracy and completeness of data can be assessed within a privacy-preserving framework. This work was however not specific to privacy-preserving data matching.

While the protocols discussed in this chapter were mostly aimed at scenarios where two or three parties are involved in the matching, further challenges will arise when more than three parties aim to match their databases in a privacy-preserving fashion, as is for example required for the first scenario given in Sect. 8.2. The problem of collusion between any pair of parties, or even a group of parties, with the objective to learn about another party's data will need to be considered carefully in such multi-party matching scenarios.

8.5 Further Reading

A good starting point to learn more about privacy aspects of data matching and to learn about different privacy-preserving data matching techniques is to read the recent reviews by Trepetin [255], Verykios et al. [262], Karakasidis and Verykios [158], and Durham et al. [99]. These reviews all discuss the advantages and drawbacks of different privacy-preserving data matching techniques. Clifton et al. [79], Fienberg [110] and Christen [58, 63] discuss various real-world data matching scenarios that illustrate different issues with regard to privacy and data matching.

For application of data matching in the biomedical area, and issues relevant to privacy and confidentiality that need to be considered, the interested reader is referred to the articles by Kelman et al. [161], Churches [73], Rushton et al. [228], Chaytor et al. [53], Durham et al. [98], Malin et al. [182], and Gliklich and Dreyer [122]. Narayanan and Shmatikov recently provided some general insight into what personally identifying information is, and myth and fallacies surrounding such information [194].

There is currently no book dedicated to privacy-preserving data matching, however there are several books on the related topic of privacy-preserving data mining [4, 257]. Many techniques developed in privacy-preserving data mining are also employed in privacy-preserving data matching.

Also related to privacy-preserving data matching are techniques that allow secure querying of a database in such ways that the database owner does not learn which records were matched with a given query. Du et al. [96] investigated this topic for different scenarios, including exact and approximate pattern matching. Similar work has also been conducted in the area of private information retrieval [55].

Chapter 9
Further Topics and Research Directions

9.1 Geocode Matching

Geocode matching, also known as *geocoding*, is the process of matching geographical information (such as addresses, postcodes or zipcodes, or points of interests) to geographical locations. These locations are commonly expressed as coordinates (latitude and longitude), which correspond to a point or area on the Earth's surface. Records that have geocodes attached to them can then be loaded into geographical information systems (GIS) and used for spatial data analysis. Geocoded records can also be inserted into the metadata of multimedia files, such as images or videos, a process known as 'geotagging'. An analysis conducted by the US Federal Geographic Data Committee estimated that between 80 and 90 % of all governmental databases contain geographical location details in some form or another [256]. In most cases, these locations correspond to the addresses of individuals or businesses, or points of interest for certain applications.

Geocoding has been employed in many application areas for quite some time. In the health sector, for example, the geocoding of patient databases has been used to analyse local clusters of diseases, or to visualise where certain disease cases occur. Rushton et al. [228] have recently reviewed how geocoding can be employed in cancer research to identify geographical correlations between cancer cases and nearby environmental factors (such as chemical factories, or sources of radiation) that might influence the occurrence of certain types of cancers. Geocode matching is also used by businesses to better understand where their customers live, and for example where to open new stores. Similarly, national statistical agencies use geocoding to assign households into local statistical areas (or census areas). These areas are often the basis of statistical analyses, and they are used by governments to plan where new facilities such as schools, hospitals, shopping centres or roads are required to deal with future population growth.

Compared to general data matching, the process of geocode matching has several special characteristics that need to be considered. The first of these is in what form the location data are available, because this will dictate how the geocode matching

P. Christen, *Data Matching*, Data-Centric Systems and Applications,
DOI: 10.1007/978-3-642-31164-2_9, © Springer-Verlag Berlin Heidelberg 2012

Fig. 9.1 An example of
the two main geocoding
techniques: Property centre
based (circles numbered 1–
9) and street segment based
(the thick dark lines and dots
numbered 10–17), with the
dotted lines corresponding to
a global street offset. Adapted
from [63]

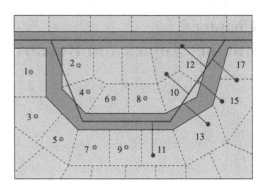

can be conducted. As Fig. 9.1 illustrates, there are two basic approaches to geocode
matching. In the first, a database containing cleaned and standardised addresses and
their geographical locations (the latitudes and longitudes of the centres of properties)
is available that covers a region or even a country [67]. For such property centre or
'property parcel' based geocode matching, ideally a database contains all known
addresses in a certain geographical area. The second approach is to use a street-
centreline database, which is made of the geographical locations of small street
segments. When an address is matched to such a street segment, its geographical
location is extrapolated based on the start and end locations of the street segment
and the corresponding start and end street numbers, as well as a street offset (usually
in the range of 10–20 m). As Fig. 9.1 shows, the street-centreline based approach
can lead to locations being extrapolated into the wrong area (for example a different
property), as is the case with property numbers 10, 12 and 17.

Depending upon the quality and coverage of the reference database used for
geocoding (either property parcel or street-centreline based), the resulting matching
quality can be quite different, as several studies have shown [48, 67, 246]. Property
parcel based geocode matching often leads to higher quality geocoding results, especially in rural areas where properties are located at irregular intervals, and where
street segments can be of lower accuracy compared to urban areas [48]. Even a small
difference in a calculated location can lead to a record being inserted into the wrong
local statistical area, and this can have significant implications for the results of any
follow-up data analysis that is based on these local statistical areas [67].

The actual geocode matching process is generally based on detailed address information, such as street number, street name, and street type; and town or suburb name,
postcode or zipcode, and state or territory name. The general data quality issues that
have been discussed in Chap. 3, and especially in Sect. 3.2 on p. 44, that are relevant
to geocoding include missing address details, typographical errors and variations in
street and town names, and out-of-date addresses. Depending upon how address data
have been entered (scanned from handwritten forms, dictated over the telephone,
or hand typed), a geocode matching rate of around 70 % is sometimes seen to be
acceptable [209].

A difference to general data matching is that geocode matching can be accomplished using a hierarchical approach [67]. Ideally, a given address in a user database is matched exactly to an address in the reference database, leading to an exact geocode location. This corresponds to an exact match in general data matching. Even with small variations in the user address (such as a typographical variation in a street name), an exact address match can often be found. However, if a user address does not include a street number, or if a given street number does not exist in the reference database, then either only an extrapolated location (between the nearest existing addresses) or the 'centre' location of the street can be assigned to that address. If even the street address cannot be matched, then only the centre of the postcode or suburb area can be assigned to the user address. An alternative to assigning the 'centre' of a street, postcode or suburb as the location to a record (which is often quite meaningless) is to assign a bounding box which either includes the street area of the given user address, or its full postcode or suburb area. Such a bounding box consists of the locations of four points that completely include either a street or even a postcode or suburb area.

As has previously been discussed in Sect. 8.2, geocode matching can have implications with regard to privacy and confidentiality of the data that is being geocoded, because knowing the locations of where cancer patients live, for example, might reveal their identity. The issues of geocoding techniques with regard to personal privacy have been discussed by Armstrong and Ruggles [13]. A topic of more recent interest that also has implications with regard to the privacy of individuals is 'reverse geocoding', the technique of assigning an address to a point on a map [45, 46, 87, 268]. With the increased public availability of digital maps that show details of, for example, certain crimes or disease cases, it is becoming easier to find out who lives at the addresses that correspond to these locations. Further work is required to develop both technologies as well as regulations that assure private information cannot be identified through such reverse geocoding techniques.

9.2 Matching Unstructured and Complex Data

Traditional data matching techniques commonly assume that the data to be matched consists of records that are stored in flat database tables, spreadsheets or text files (such as comma separated values (CSV) or tabulator separated values). Each such record refers to one entity, it contains the same attributes, and there is no hierarchical structure between the attributes in these records. Commonly, the data to be matched also only contains one type of entity.

Modern information systems, however, consist of data that are stored in formats different from such single database tables. Relational database systems are generally normalised, such that different components are stored in different tables that are linked with each other through unique keys. For customer or patient databases, for example, the addresses of where people live are likely stored in a table that is separate from the names and dates of birth of these individuals. Records to be used for data matching need to be extracted from such normalised databases through queries that combine the required attributes from different tables into a single table or view.

Increasingly, data are stored in repositories and formats different from relational databases. Two types of data formats that have gained popularity in recent times are Extensible Markup Language (XML) and Resource Description Framework (RDF). Many different XML-based schema languages have been developed for different application areas. An advantage of data stored as XML documents (in text format) is that they can easily be interchanged between different applications that support parsing of XML documents. A schema language specifies the set of elements that can be included in an XML document for that language, the relationship between these elements, and what attributes can be applied to them. Schema languages are described either as Document Type Definition (DTD) or XML Schema. XML documents can be represented in a tree structure, where elements can consist of subelements. For example, a person element can contain a name and an address element. The name element might contain subelements such as title, given name, middle name, initials and surname, while the address element contains many subelements that correspond to address components such as those shown in Fig. 3.1 on p. 42.

Weis and Naumann developed techniques to exploit the hierarchical structure of XML documents to improve the detection of duplicates in XML databases [195, 269, 270]. Specifically, in their approaches the comparison of XML documents proceeds in a top-down approach, such that documents that do not match at a certain level in the tree structure are not compared further [195]. The idea behind this approach is that, for example, if the country values of two XML records are different, then even if the city values are the same the records are unlikely to be representing the same entity. This makes intuitive sense, in that cities with the same name can occur in different countries. There is, for example a 'Melbourne' in Florida, USA, as well as in Australia, towns named 'Newcastle' are located in the UK, the USA (at least six of them) and Australia, and there is a town called 'Springfield' in nearly every English speaking country.

The RDF format to store data has recently gained a lot of interest from the semantic Web and linked data communities. The RDF data format consists of triples made of a subject, a predicate, and an object. The subject refers to a resource, usually a Uniform Resource Identifier (URI); the predicate denotes an aspect of the resource and also describes the relationship between the subject and the object. As a simple example, assume one wants to store the given name 'peter' and surname 'miller' of a record with identifier 'a1' into two RDF triples. The subject for both triples is 'a1', the predicate for the first triple is 'has given name' and its object is 'peter', while the predicate for the second triple is 'has surname' and its object is 'miller'. Volz et al. [264] and Jentzsch et al. [150] describe a data matching framework that can match data stored as RDF tuples, while allowing the specification of indexing and approximate string comparison functions through XML documents. This framework is further described in Sect. 10.2.10.

For data that are stored in free text format without any specific attributes, such as data extracted from Web pages, emails or news articles, information extraction techniques, such as parsing and segmentation [230], generally need to be applied before such data can be used for data matching applications. As was discussed in Chap. 3, the quality of such data pre-processing is crucial in order to achieve high quality matched data.

Yakout et al. proposed a data matching approach not on databases that contain records that refer to entities, but rather on databases where each entity is represented by a set of transactions, each consisting of a time-stamp and a set of features [291]. These transactions can for example correspond to market baskets consisting of items a customer bought in a shop. The matching is based on extracting patterns (such as the weekly shopping behaviour of customers). For each entity, a behaviour matrix is extracted from the transactional database that characterises which actions (such as items bought) were performed by an entity at what time (or on which day). These matrices are then further transformed into a condensed behaviour representation, and a similarity score is calculated between pairs of entities based on their condensed representation. Experiments by the authors using a real-world data set sourced from the Walmart retailer showed the feasibility of their proposed approach on using transactional data for data matching [291].

9.3 Real-time Data Matching

Much of recent research in data matching has concentrated on the development of sophisticated classification techniques that improve data matching quality compared to the basic threshold-based probabilistic approach, as was discussed in Chap. 6. In practical applications of data matching, however, accurate matching is only one of several issues that must be addressed. Commonly, practical applications will need to involve a trade-off between matching accuracy and qualities such as matching speed, scalability to very large databases, and increasingly, real-time matching capabilities.

Most existing data matching and deduplication techniques are aimed at the offline batch processing of static databases. Many organisations, however, are challenged by the task of having a stream of query records that contains information about entities that need to be matched to one or several databases of known entities.

An example of real-time data matching can be found in the area of consumer credit applications, where a database that contains information about credit cards, loans and other consumer credit details is being queried with records that correspond to credit checks, identity checks, changes in the personal details of existing consumers, or new customers that are being added [212]. These query records will be supplied by different customers, such as banks and other financial institutions. The organisation which holds this consumer credit database (sometimes called a 'credit bureau') must be able to provide responses to queries in (near) real time, ideally within a few seconds at most.

Similar real-time data matching requirements can be found in online government services (where the identity of a citizen needs to be verified in real time), or in law enforcement or national security database systems where police officers or intelligence agents need to be able to find the details of suspects in real time while at the same time allowing approximate matching of personal details [54, 91, 267].

Real-time data matching has much in common with the functionality of text and Web search engines [21, 43, 303]: real-time and approximate matching and ranking of results. However, the databases upon which data matching is commonly applied do not contain documents that provide a rich feature space. Rather, these databases are made of structured records with well-defined attributes that often only contain short strings or numbers, such as the names or other personal details of people.

The first approach to 'query-time' data matching was proposed by Bhattacharya and Getoor [32] based on a collective classification approach, which was presented in Sect. 6.10. The idea of this approach is to conduct the matching for each query record in the raw 'unresolved' database that might contain duplicate entity records. While this approach can improve the matching quality significantly compared to simple pair-wise matching on the same database, due to the complexity of the collective classification approach the matching was not feasible in real time. An average matching time of 31 s per query record was reported when this approach was applied on a database consisting of 831, 991 records [32].

Christen et al. approached the challenge of real-time matching from an information retrieval angle by pre-calculating the similarities between attribute values and storing them into a set of specialised data structures [69, 70]. At query time, these pre-calculated similarity values can be efficiently retrieved, and the overall similarity between the values in a query record and the relevant database records is calculated. The data structures employed consist of one inverted index [288], which contains the blocked attribute values (using standard blocking as was described in Sect. 4.4); a second data structure in the form of another inverted index which contains attribute values as keys and lists of other attribute values they are similar to; and a third inverted index data structure which for each unique attribute value consists of a list of all record identifiers that contain this value. The experimental results reported by the authors showed that such an index-based approach can reduce the matching time by over two orders of magnitude compared to a traditional matching technique [69, 70]. The reported matching time for a query record on a database with nearly 7 million entity records was below 0.1 s in average.

More recently, Ioannou et al. proposed an entity-aware query processing system which works on probabilistic databases where each record is assigned an uncertainty probability that is used in the matching process [148]. Similar to the work by Christen et al. [69, 70], possible matches (which have a likelihood weight attached to them) between records are stored in a database, and the attributes of all records have themselves likelihoods attached. These likelihoods reflect the confidence one has in the correctness of the attribute values given in a record. Lower likelihood values can for example be assigned to records which have been sourced from untrusted databases, or which contain inconsistent values. When a query record is given to such a probabilistic database, a matching and merging step of entity records is conducted based on the given likelihood values, and all merged entity records that fulfil the query terms are being returned to the user. In order to achieve efficient and fast query-time processing, a dynamic index data structure is maintained, which contains subsets of entities that are connected with each other. This data structure is updated as entity records are added or removed (a topic that will be further discussed in the following section). The authors reported matching times of around 70 ms per query record on databases consisting of around 50,000 entity records [148].

The latest work relevant to real-time data matching is an approach proposed by Dey et al. who developed a matching tree for efficient online data matching [91]. Their basic idea is to limit the amount of communication required when records are matched between disparate databases by re-ordering the comparisons that are conducted between record attributes, such that a match or non-match decision can be made as quickly as possible without the need to compare all attribute values between records. Their approach is a modification of the traditional threshold-based probabilistic record linkage technique that was described in Sect. 6.3. The authors illustrate the significance of their approach through two example applications, the first coming from the area of insurance claim processing, where health insurance claims sometimes have to be processed by several insurers, and data matching is required to make sure no double billing is happening. The second application is in the area of crime investigations, where police officers in the field need to be able to query databases from different municipality in order to identify a potential suspect. Real-time and approximate matching is required in this application, and the amount of communication of such queries should be as small as possible. The matching tree developed by Dey et al. is trained in an offline phase from both matching and non-matching examples, such that in the online (query) phase a match or non-match decision can be made quickly. Experiments conducted by the authors showed that their approach can significantly reduce the communication overhead without any loss in matching quality [91].

9.4 Matching Dynamic Databases

Related to real-time matching is the issue of dynamic databases. As the example of consumer credit applications in the previous section has shown, modern information systems do not simply consist of static databases, but rather these databases are modified on an ongoing basis. These modifications can either be inserts of records that correspond to new entities (new customers, babies born, new students, new employees, and so on); the removal (or inactivation) of old records that correspond to entities that are not relevant anymore for a certain application (such as people who have died); and updates of the personal details (like name or address changes) or other application specific details (like student enrolment status or employment level) of entity records. In many applications, the latter category of modifications is the most frequently occurring one.

Dynamic databases require that all steps of the data matching process can handle modifications. While data pre-processing can be conducted independently for every record before it is added into a database or used to update an existing record, the indexing, comparison, and classification steps need to be modified. The data structures used for indexing (often based on inverted indexes [288, 303]) must facilitate that new values can be added, old values can be removed, and that weights stored in these indexes (such as term-frequencies and document frequencies [288, 303]) can be modified efficiently. The indexing technique used in real-time data matching will determine which new records will be compared with records that are already stored in the entity database. The same comparison functions as used with standard static data matching can therefore be employed to compare records. However, for certain applications it might be advisable to modify the calculated similarities between records according to the time when these records were added to a database or when they were last modified. Such temporal information is captured by a time-stamp for each record in a database.

For example, when addresses are compared, depending upon the application, it might be sensible to discount matches with addresses that are more than a few years old. This is because the longer ago an address was recorded for an individual the higher the likelihood is that the person since then has moved to a different address. Discounting matches with older records, similar as for example is done in data stream processing [179], is one possible approach on how temporal aspects can be taken into account. It might, for example, be more appropriate to have a query record match with a more recent database record rather than more older ones, even though older records have higher similarities with the query record.

The classification step, finally, requires that the classification model can be dynamically updated as new records are added into a database or existing records are modified. The alternative of re-calculating (periodically) from scratch a classification model might be too time consuming and also lead to a drop in matching accuracy as the classification model gets older, an issue known as concept drift [136].

Research into data matching on dynamic databases has only recently started to gain attention. The work by Ioannou et al. [148], which was already discussed in the previous section, allows dynamic updates of the data structures used as new records are added into the probabilistic database on which this approach is based upon.

Whang and Garcia-Molina developed an approach to data matching that allows matching rules to evolve over time as new data become available [274]. Rather than having to conduct a full data matching process from scratch each time new records are added into a database, their approach is based on materialised matching results in the form of sets of records that are classified as matches. It is assumed that matching rules (as was discussed in Sect. 6.5) are provided by a user who over time refines these rules, for example because they are not happy with the results of a data matching, or because data with new characteristics are added to the databases which requires modified rules. The authors propose efficient algorithms for clustering based data matching classifiers using either match-based clustering or distance-based clustering. An experimental evaluation on real shopping and travel data showed that their approach can speed-up data matching when rules evolve compared to the naive restarting of the full data matching process [274].

Li et al. more recently investigated how temporal information available in the databases to be matched can improve the matching quality [179]. They specifically investigated bibliographic data where the dates of publications are included, and how this information can be used. In their data, each author is affiliated with a certain institution over a period of time. This information is used in the similarity calculation between records. Additionally, their approach modifies similarities between records according to a time decay, where candidate record pairs that have larger differences between their time-stamps have their dissimilarities penalised less because it is more likely that values change over time (such as the affiliations of authors). On the other hand, agreements are more penalised over longer time differences as it is less likely that the same entity will keep the same value over a long time period. While their approach does not directly address the issue of dynamically changing data (they assume the data to be matched are static but each record contains a time-stamp), the idea of decay rates for similarities and dissimilarities could also be applied for dynamic databases, by adaptively adjusting decay rates according to new data that are added to the databases to be matched.

The approach proposed by Yakout et al. [291] on conducting data matching based on transactional records that correspond to the behaviour of entities rather than the actual entity records, as was already discussed in Sect. 9.2, also takes temporal aspects of transactions into account. While their approach is not specifically aimed at dynamic databases, similar to the work by Li et al. [179] described above, an extension to this behavioural data matching approach would allow the dynamic adaption of the calculations of similarity scores between sets of transactions as new data become available.

9.5 Parallel and Distributed Data Matching

As modern databases are becoming larger, deduplicating or matching them requires increasingly large amounts of computing power and storage resources. Researchers have begun to investigate how modern parallel and distributed computing environments can be employed to reduce the time required to conduct large-scale data matching projects [25, 66, 88, 160, 163, 165]. Several approaches to parallel data matching aimed at different computing platforms (such as multi-core machines or distributed processors) and parallel programming environments (such as the Message Passing Interface (MPI) or Map-Reduce) have been investigated.

The major distinctions in parallel architectures are (1) if all processors have access to the same shared memory or not; (2) if all processors have access to the same file system or not (shared or distributed I/O); (3) if the same program runs on all processors (data parallelism) applied on different subsets of the data, or if different programs run on different processors (task parallelism), either applied on all data or on different subsets of the data; and (4) how the processors are connected with each other (network topology) [126]. Parallel programs are either implemented in a master-worker style, where a master process assigns tasks to worker processes; or in a single-program multiple-data (SPMD) style where all processors run the same program on different subsets of the data, but no single process controls the overall parallel execution.

Load balancing is a crucial aspect that needs to be carefully considered in order to achieve scalable parallel programs. All processors of a parallel computing platform need to be occupied all the time, as otherwise one processor might finish its task before others and therefore potentially has to wait for other processors to finish before a next computation phase can be started. Good load balance can be difficult to achieve for applications where the distribution of data to processors depends upon the actual values of the data, which can be irregular or skewed. As will be further discussed below, this is often the case for data matching applications.

Parallel performance is generally measured as the speedup in run-time achieved when more than one processor is employed, t_p, compared to the run-time of the same program on a single processor, t_s. Speedup is calculated as $s = t_s/t_p$. For example, if the matching of two databases on one processor takes 45 min and 12 min when four processors are used, then the speedup is $s = 45/12 = 3.75$. Ideally, speedup should be as close to the number of processors used, especially as the number of processors is increased. In order to achieve a speedup that scales linearly with the number of processors that are being used, all components of a program need to be parallelisable efficiently. If a program contains a component which cannot be parallelised, then this will severely impact on the speedup that can be achieved. Such components can consist of the program initialisation, the loading of data at one processor only, or the communication between the processors to exchange intermediate results.

For example, assume 10 % of the time required by a data matching program cannot be parallelised (denoted with t_{seq}) and 90 % can (denoted with t_{par}). The parallel run-time when using p processors can then be calculated as $t_p = t_{seq} + t_{par}/p$. Assuming the sequential ($p = 1$) execution of this program takes 45 min, with $p = 2$ processors the parallel execution time (with $t_{seq} = 4.5$ min) is reduced to $t_p = 4.5 + 40.5/2 = 24.75$ min (speedup $s = 45/24.75 = 1.82$). When $p = 5$ processors are used, $t_p = 4.5 + 40.5/5 = 12.6$ min (speedup $s = 3.57$), and with $p = 10$ processors $t_p = 4.5 + 40.5/10 = 8.55$ min (speedup $s = 5.26$). Even with $p = 100$, the parallel run-time would be $t_p = 4.9$ min and the speedup would only be $s = 9.17$. As this example shows, in order to achieve a parallel program that is scalable to a number of processors, all components of the program need to be parallelisable efficiently.

In order to achieve scalable speedups for data matching and deduplication systems, it is necessary that each step of the data matching process (as was illustrated in Fig. 2.1 on p. 24) needs to be parallelisable. The following list discusses this requirement for each of these steps:

- *Data Pre-Processing*: In this first step, each record can generally be processed independently from all others. Given that data pre-processing relies heavily upon dictionaries and look-up files that contain spelling variations, nickname expansions and so on, each processor must be able to access these files. This can be accomplished either via direct access to these files, or by communicating the required information from one processor (such as a master process) to all other processors via broadcast communication. Assuming the time required to pre-process is the same for each database record, a uniform distribution of records onto processors can achieve an evenly balanced workload on all processors (i.e. each of the p processors will pre-process n/p of the n database records).

- *Indexing*: As was discussed in Chap. 4, indexing is a crucial step to reduce the number of record comparisons that need to be conducted. The sizes of the blocks or clusters generated by most indexing techniques depend upon the distribution of the actual attribute values in the records, which for certain attributes can be highly skewed (such as for surnames or given names). The major challenge in parallelising an indexing technique is therefore to achieve good load balancing across all processors. This can be achieved by splitting the databases to be matched into blocks or clusters that will lead to an equal number of record pair comparisons to be conducted on each processor. The second issue in the indexing step is how the databases to be matched are accessed (i.e. can all processors read the full database or only a portion of it), and how candidate record pairs are being communicated between processors to achieve the required balanced workload.

- *Field and Record Comparison*: The comparison of each individual pair of records can be accomplished independently from all others. If the distribution of record pairs done in the indexing step has led to a balanced distribution of candidate pairs, then the comparison step can be efficiently conducted in parallel assuming that all processors have access to the actual attribute values of the records they are comparing.

- *Classification*: When a classification technique for data matching is being parallelised, the major concern is if either each compared record pair is classified independently from all others, or if some form of clustering or collective classification technique is applied (as was discussed in Chap. 6). For the classification of individual record pairs, each processor only needs to know the classification model and its parameter settings (such as thresholds or rules). It can then classify individual pairs and either store the classification results into a file that is being merged at the end of the classification step, or sent as a message to a coordinating process (such as the master process).

  On the other hand, the parallelisation of classification techniques that are based on groups or clusters of records, that generate global graphs of possibly matching records, or that consider the transitive closure of record groups (as was covered in Sects. 6.8–6.11) requires more sophisticated approaches. Research conducted in parallel data mining [297], such as parallel clustering algorithms, can be applied for parallel classification in data matching. Such algorithms commonly include complex communication patterns, because intermediate results often need to be exchanged between processors that hold certain subsets of the data (the compared candidate record pairs in the case of data matching).

- *Evaluation*: The final evaluation of the quality and complexity of a data matching exercise generally needs to be conducted on a single processor in order to obtain the overall final result. Each processor can however calculate these measures on the portion of the data they hold, and send the results to a coordinating process which aggregates all individual results.

This final step will also require that the actual result files (that contain the matched record pairs or groups) are collected from the different processors and merged into one single result file that can be used further.

Hernandez and Stolfo [140] in 1995 presented a parallel version of their sorted neighbourhood indexing approach (which was discussed in Sect. 4.5). Using a parallel machine where memory was not shared between processors, they employed a master-work approach where the master process distributes the database, the worker processes sort their subset of the data, and the master process joins the sorted subsets. Experimental results on up to six processors showed a sublinear speedup due to the requirement of the master process to read, distribute and collect the database subsets.

A description of a data matching system that uses data parallelism was provided by Christen et al. [66]. This system employed the Message Passing Interface (MPI) library for communication between processors in a distributed environment where each processor only has access to its own memory and disc. Experimental results confirmed that it is more difficult to achieve scalable speedups for the indexing step, however the comparison and pair-wise classification steps can achieve nearly optimal speedups. The majority of time, between 94 and 98 %, in the experiments conducted was spent in the comparison and classification steps, resulting in a nearly linear overall speedup [66].

Benjelloun et al. [25] proposed a family of algorithms based on the Swoosh algorithm (which was presented in Sect. 6.12) that facilitate distributed data matching. D-Swoosh, as it is called, is using a generic 'scoop' function which determines how records are mapped onto processors, and a 'resp' function which designates which processor is responsible for a certain record. Each processor runs the R-Swoosh [26] algorithm on its subset of the data. Each matched and merged set of records is communicated to all processors that hold a record that is affected by a merge. Different strategies for choosing the 'scoop' and 'resp' functions were considered, and an experimental evaluation was presented using a real comparison shopping data set. These experiments showed that D-Swoosh with certain parameter settings was able to achieve a nearly linear speedup on a parallel platform consisting of 16 processors [25]. A master-worker approach of D-Swoosh, called P-Swoosh, was implemented by Kawai et al. [160]. Different ways of how matching record sets are distributed onto processors (and in certain cases replicated) were explored. Two different load balancing techniques were investigated, one with a static distribution and the other with an adaptive distribution. The speedup results of experiments, that were using the same data set that was used by Benjelloun et al. [25], were comparable to the results achieved with the D-Swoosh approach.

Kim and Lee [163] investigated different approaches to match and merge data sets that are either clean (i.e. the two source data sets do not contain duplicate records) or dirty (i.e. the source data sets can contain duplicate records). The general idea of their approach is that the smaller of the two data sets to be matched, **A**, is replicated on each processor. The larger data set, **B**, is however distributed. In the first step, each processor conducts the matching and merging of records between **A** and its subset of **B**. In the second step, the results of the first step are exchanged between processors as required, and a matching and merging of the exchanged records is conducted. The second step is repeated until no more merges can be conducted. In order to achieve good load balancing, the number of record pair comparisons that need to be conducted is estimated based on the size of the data stored at a processor. This estimate is used to distribute the actual data across processors. The proposed algorithm was implemented in a distributed Matlab environment. Experiments on synthetic bibliographic data showed that this approach can achieve speedups of up to 7.5 on 8 processors, and that it is between 11 and 18 % more efficient than P-Swoosh [163].

More recently, Kirsten et al. proposed different strategies of how to distribute data to generate independent matching tasks that can be executed on different processors [165]. This data parallelism approach specifically aims to set the number of partitions and their sizes in an optimal way with regard to both communication overhead and memory requirements. A master-worker program style was adapted where the master processor (called the workflow service) calculates the number and sizes of partitions which are then distributed to the workers (called the match services) by a data service. This data service also collects the individual results from all match services. Two partitioning approaches are proposed. The first is based on the full Cartesian product of all possible record pairs that can be generated from the two databases that are being matched (i.e. no indexing is applied). The second partitioning approach assumes that an indexing technique is used (this partitioning is independent of the actual indexing technique employed). Because the size of blocks depends on the data to be matched, the indexing based partitioning approach splits large blocks into smaller ones and merges small blocks into larger ones in order to achieve better load balancing between the different processors. Experimental results using a real data set of consumer products from a comparison shopping Website on a parallel platform with 16 processors showed that good speedup results can be achieved by both partitioning approaches [165].

A data parallelism approach for multi-core computing platforms based on the Map-Reduce programming model was recently presented by Dal Bianco et al. [88]. They proposed an efficient indexing approach based on standard blocking that, similar to the approach presented by Kirsten et al. [165], in the first step generates larger blocks that have lower similarities using a less specific blocking key definition. Because such large blocks can lead to imbalanced workloads, in the second step the blocks that are too large according to some criteria are split into smaller blocks using a more specific blocking key definition. After this indexing step, the generated and compared record pairs on each processor are formed from the records stored on that processor, and the results are merged into a single output file. Synthetic data sets consisting of up to 4 million records that were created with the FEBRL data generator (described in Sect. 7.6) were used to evaluate the proposed approach. Speedup results of between 3 and 3.5 were reported on a multi-core machine with four processors [88].

Kolb et al. [167] recently proposed a similar approach to parallel data matching also based on the Map-Reduce programming model. However, instead of using the standard blocking technique they developed two efficient parallel variations of the sorted neighbourhood indexing technique that was discussed in Sect. 4.5. In the first approach, multiple Map-Reduce processes are employed, while the second approach employed a tailored data replication to improve performance. A pre-processing step was applied on the data to be matched that calculates a matrix that contains information about how many entities are given per blocking key. Based on this information an efficient load balancing of entities onto processors can be achieved. A data set containing around 1.4 million bibliographic records was used to evaluate the proposed approach, and speedup results of around 7 on 8 processors were reported [167].

9.6 Research Challenges and Directions

This section provides a discussion of current challenges and directions for future research in the area of data matching. The topics listed have been compiled with the help of some of the world's leading data matching researchers and practitioners. The aim of this list is to provide researchers and graduate students who are entering the area of data matching with some ideas for potential research topics. The interested reader is also referred to Winkler's report on research directions published in 2006 [284]. Most of the topics described in his report are still valid in 2012.

The given list is ordered following the structure of the book, i.e. starting with topics that refer to the different steps of the data matching process, and ending with topics related to privacy and the topics discussed earlier in this chapter.

- *A unifying framework for data matching*: While a variety of data matching prototype systems has been developed in recent years [168], it is currently not possible to easily compare these systems because they employ different methodologies and evaluation measures, and different test data sets have been used for their evaluation. What is required is a unifying framework for data matching that allows the integration of different algorithms such that comparative evaluations can be conducted more easily. Such frameworks must allow the different functionalities of the data matching process (implemented as methods, functions or modules) to be exchanged between prototypes, and facilitate that new algorithms and techniques can be plugged into existing systems. Frameworks in the areas of data mining and machine learning, such as the WEKA toolbox [132], have shown to be very successful both for research and education in these areas.
- *A specification language for data matching*: Related to the previous topic is the desire to have a language that allows specifications, models and parameters for all steps of the data matching process to be written in an implementation independent way. Such a language will allow the exchange of specifications across different data matching systems, facilitate comparative evaluation, and improve portability. In other domains, such specification languages have attracted interest both from academia and industry. One example is the Predictive Model

Markup Language (PMML)[1] developed by the Data Mining Group. PMML allows cross-platform data mining model development and deployment. A standardised specification language for data matching will help to bridge the gap between data matching research and commercial products, because it will facilitate for example the exchange of practical data matching specifications from industry into academia, thereby making academic research more relevant for solving real-world data matching problems.

- *Data pre-processing to achieve a certain match quality*: When the quality of data is assessed for the first time prior to their use in data matching, it is generally not known how much efforts in data pre-processing (cleaning, standardisation and segmentation) are required to convert the input data into a form that will lead to data matching results of high quality. Given data pre-processing is a time-consuming and often labour intensive process, techniques are required that provide an estimate of how much effort should be spent on data pre-processing in order to achieve data matching results of a certain quality. For some data matching applications extensive data pre-processing might not even be required.

- *Incremental data pre-processing*: Related to the previous topic is the question of how data pre-processing can be conducted incrementally, for example if the outcomes of a data matching exercise require further data pre-processing to be conducted in order to improve matching quality. For certain applications, an iterative approach to data pre-processing might be appropriate, where the results of the first matching iteration are used to conduct further data pre-processing, which in turn can lead to improved matching quality in the second iteration, and so on. Such an approach that conducts the two steps data pre-processing and data matching iteratively until a certain match quality is achieved will require techniques that allow dynamic adjustments of data pre-processing and data matching algorithms and their parameters. Ideally, these parameters are learnt automatically throughout this process.

- *Domain specific data pre-processing*: Because most data pre-processing is domain specific, techniques are required that take domain knowledge into account both to improve the quality of the pre-processed data as well as to reduce the manual efforts required in the data pre-processing process. Domain knowledge needs to be exploited and represented such that it can easily be reused and modified when different data are to be pre-processed. One way to achieve this could be to encapsulate domain knowledge into a specification language that can be used by different data matching systems.

- *Using external data sources for automatic data pre-processing*: With the availability of an increasingly large and diverse amount of information from different data sources, the question is how can data quality be improved through techniques that automatically (or semi- automatically) exploit such external data. Because the quality of external data might be questionable, the confidence one has in such data sources needs to be taken into account during the data pre-processing process.

---

[1] http://www.dmg.org/

- *Handling missing data*: How to efficiently deal with missing data is a fundamental question in data matching. There are different reasons why missing data can occur. Some missing values might refer to truly unknown or non-existing values (such as a person who does not have a middle name), while others might be due to data entry errors, equipment malfunctioning, or simply because it was decided a certain value does not need to be recorded. Depending upon the reason why missing data occur, the parameters used in data matching algorithms might have to be modified for individual records. A question is if it is known why a certain record has a missing value or not (for example from meta-data that provides details about the data entry conditions). Such information could be used in the comparison and classification steps to adjust the similarity calculated between attribute values, or to decide the match status of a candidate record pair.

- *Estimating match and non-match rates*: Because in many data matching situations no ground-truth data are available, it can be challenging to estimate the accuracy of record pairs that have been classified as matches or non-matches. Statisticians have worked on the problem of estimating match and non-match rates for many years [23, 283, 285]. Because the candidate record pairs generated depend upon the indexing technique used in a data matching exercise, and their classification is influenced by both the comparison functions and the classification technique used, the assessment of match rates needs to take all these techniques and their parameters into account. If no ground-truth data are available, data from external sources, earlier matching projects on the same data, or synthetic data that have similar characteristics as the real data that are to be matched, can potentially be used to estimate match and non-match rates. Sampling of candidate record pairs and their manual evaluation seems to be a viable alternative [283]. Further research is required into how such estimates can be used with the advanced classification techniques for data matching that have been developed in recent times (such as collective classification).

- *Adjusting statistical analyses for matching errors*: When matched data are used for further statistical analysis, it is important to know the error rate that has been introduced in the matching process. However, as was discussed in the previous topic, an exact value of this rate is often not known and needs to be estimated. Based on such estimates, any follow-up statistical analysis needs to be adjusted accordingly. Work in this area has again been conducted by statisticians [50, 174, 235], however it seems the computer science community, for example researchers in the fields of data mining and machine learning, has so far not investigated this topic. In these fields, it is generally assumed that a matched data set is accurate and complete. This means the records in a matched data set correspond to correct matches, and that the matched data set contains all true matches between the two source data sets.

- *Improved classification*: Current classification techniques for data matching are generally based on the similarity values obtained from the comparison functions when records are compared with each other, and from additional relational information which links records with certain common characteristics with each other. Classification techniques that select the best available features to make accurate match and non-match decisions could help to improve the match quality that can be achieved, while feature selection techniques can help to decide which features (i.e. which comparison functions applied on which attributes) have the highest discriminating power to distinguish between matches and non-matches. Novel techniques that allow classifications to be made as quickly as possible without the need to compare all attributes between two records have already helped to improve the performance of data matching systems [91, 193].

Extending data matching from relational data to more complex types of data (as was already discussed in Sect. 9.2) will require novel comparison as well as classification techniques that can integrate different types of data, including text, images, video, as well as spatial (such as geo-referenced) and temporal data.

- *Incremental data matching*:As was previously discussed in Sect. 9.4, there is a move away from matching static databases towards online and real-time data matching systems, where the underlying databases are constantly modified with new and updated records. Data matching techniques applied on such databases must be able to deal with the dynamic nature of these data. It must be feasible to dynamically update the classification models used, for example by adjusting parameters or updating rules [275]. How such model adjustments can be best achieved in a running environment while keeping high matching performance is an open research question.

- *Semantic matching*: Current data matching techniques are mostly based on the syntactic matching of attribute values that are available in the databases that are matched, and potentially by using additional relational information between records. Semantic matching [299] requires techniques that can detect which attribute(s) in one database correspond(s) to which attribute(s) in another database (similar to schema matching techniques [224]). Because large real-world database systems can be very complex and consist of hundreds of tables, manually determining which tables correspond to each other will likely be very expensive. Semi-automatic techniques that explore correspondences and correlations between tables and their attributes are required [24].

- *Incorporating user knowledge*: With many commercial data matching systems being either based on rules or ad hoc implementations that have been hand-crafted and improved over long periods of time, it is important that the human knowledge and expertise encapsulated in these systems are not lost (for example when an experienced data matching expert retires from an organisation). Most of this knowledge is very domain specific, and it changes over time. Some rule-based data matching systems consist of hundreds if not thousands of rules that have been carefully manually designed. For many organisations these rules are very valuable as they encapsulate important aspects of their business intelligence. It is therefore crucial that such human expertise can be captured and transformed into representable knowledge such that it can be ported to other data matching systems or even be incorporated into future systems.

- *Matching across many different data sources*: In certain applications, data from more than two sources need to be matched, for example in situations where several government departments work together to identify individuals who might have committed fraud, or where medical records of patients who potentially have been admitted to several hospitals need to be combined. Records that correspond to the same individual will likely have been recorded at different points in time, and therefore potentially have different attribute values (such as a changed address or surname). The various databases might also have different structures and formats, and the records stored in them can contain different types of information (i.e. different attributes according to the need of the application the data was collected for). The quality of the data, as well as the confidence one can have in their correctness will likely also differ.

In other applications, databases that contain different types of entities might need to be matched, for example scientific publications that are matched with authors or institutions to help assess the impact of researchers and the organisations they are working for.

When several databases are to be matched, the matches identified between records from two databases can help to inform the matching of records from other databases [276]. Even when only two databases are matched, information from external data sources, sometimes called 'bridging files' [284], can be used to find matches that are not obvious from the records in the two databases only. Such a bridging file might contain the details of the last three addresses of people and the time periods when they lived at these addresses. This information will facilitate matching between databases that were recorded at different points in time.

The techniques used in the indexing, comparison, classification, and evaluation steps of the data matching process need to be reconsidered carefully when data from more than two sources are being matched, and new efficient approaches need to be developed. Some initial work in this direction has recently been conducted by Sadinle et al. [229] who investigated how the traditional Fellegi and Sunter approach to threshold-based probabilistic classification (as was presented in Sect. 6.3) can be extended to more than two data sources.

- *Efficient clerical review methods*: In traditional data matching systems where individual pairs of records are classified into matches, non-matches and potential matches (as was discussed in Sect. 6.2), the manual clerical review of pairs that were classified as potential matches can be conducted via simple user interfaces such as the ones presented in Figs. 7.4 and 7.5 on p. 175 and 176. The problem with such a simple approach is that the context of the record pair under review is not shown to the person undertaking the review (such as how many other people in the same database have the same name). In a similar way as collective classification and clustering techniques can improve data matching quality, improved visualisation techniques for the clerical review process need to be developed. One such recent effort is the D-Dupe prototype system presented in Sect. 10.2.2 [36].

  Ideally, a clerical review system needs to visualise a pair or group of records which allows viewing them as being part of a network, connected to other records through matching attribute values and their similarities. For example, different colours of connected records could illustrate which attributes are matching between records (for example green if there is an address match, red for a name match or blue for a date of birth match), while the thickness of a connection between records in the network illustrates the similarity between them. Such improved user interfaces for clerical review will allow reviewers to make better informed decisions about the match status of pairs or groups of records. Graphical interfaces can also be used to highlight records that have been matched but where some type of contradictions in the match status with other records might occur, such as a violated transitive closure condition, as was discussed in Sect. 6.8.

- *Scalability and speed of computation*: As databases are getting larger, deduplicating and matching them requires increased computational efforts, and larger storage and memory resources. The development of novel efficient indexing techniques is crucial to make the data matching process scalable to very large databases. Ideally, the number of candidate record pairs that are generated by an indexing technique should only scale linearly with the size of the databases to be matched. As was discussed in the previous section, the development of techniques that allow the efficient parallelisation of data matching is a current research topic [25, 66, 88, 160, 163, 165, 167]. Parallel approaches that make use of modern computing platforms, such as multi-core processors or grid- and cloud-based systems, will facilitate data matching on very large databases. When external services, such as those offered by cloud computing providers, are employed for data matching, then privacy and confidentiality become crucial aspects that need to be considered. The second issue is in which step of the data matching process to best spend funding for computing services, assuming a commercial cloud service will be charging according to the amount of computing services that are required by a certain process.

  A different approach to improve the computational efficiency of data matching is to better couple data matching techniques into relational database systems, and to better exploit the mature optimisation techniques that have been developed by the database community over the past decades. Using for example an efficient nested loop join algorithm and disc based indexes that are implemented in modern databases systems could potentially lead to large performance gains.

- *Effective benchmarks and test data collections*: A major obstacle for data matching research is the lack of large test data collections and publicly available benchmark data sets that allow the comparative evaluation of data matching algorithms and methods. Researchers commonly use either their own data sets for testing, use one of the several small test data sets that are publicly available (as was discussed in Sect. 7.5), or they generate their data using one of the data generators presented in Sect. 7.6 or their own data generation program. Koepcke et al. recently discussed the issues related to evaluating data matching systems in more detail [169]. Privacy and confidentiality are the main reasons that no large real databases that contain personal information have been made publicly available for research purposes. Nevertheless, without large-scale benchmarks, the outcomes of data matching research will continue to be difficult to evaluate comparatively. As a result it will be difficult to identify significant advances in data matching technologies.

  An alternative approach to developing test data collections is to set-up online test environments where researchers can submit their data matching algorithms, which are then evaluated on a set of standardised benchmark data sets using a variety of evaluation measures. Weis et al. proposed such a system to evaluate the deduplication of XML data sets [271]. An extension of such a benchmark system to different types of data and its actual implementation and use by researchers would be a big step towards efficient and objective comparison of data matching algorithms and techniques.

- *Multi-party privacy-preserving data matching*: When data from more than two sources are to be matched and privacy and confidentiality need to be considered (for example because each data source might be located in a different organisation and the data contain sensitive information that cannot be given to other organisations in plain-text as was discussed in the first example scenario given in Sect. 8.2), then not only the issues regarding matching of more than two data sources discussed previously need to be considered, but also how such multi-party matching can be conducted while privacy is preserved. Issues that need to be considered include how collusion between (subsets of) parties can be prevented. Given the generally high computation and communication requirements of privacy-preserving data matching techniques even for matching data between two parties, new efficient protocols need to be developed.

Chapter 10
Data Matching Systems

10.1 Commercial Systems and Checklist

The number of commercial products that provide some form of data matching or
deduplication capabilities is hard to judge, because there is no clear distinction of
when a software should be judged as a data matching system or not. Certain products
offer only some functionalities related to data matching (such as approximate string
comparison functions), while others provide functions for most or even all steps
of the data matching process. Some products are small innovative stand-alone software packages or libraries, while others are large systems that are aimed at business
data integration, business intelligence, or customer relationship management. Data
matching and deduplication are only a small component of such large systems. Most
commercial database systems also provide various extensions to the standard SQL
language that allow phonetic string encoding such as Soundex or NYSIIS (as was
covered in Sect. 4.3), or approximate matching of string values using for example an
edit distance based approach (as was discussed in Sect. 5.3).

The large number of commercial systems, and the dynamic nature of the IT market,
make it difficult to provide an overview of specific commercial data matching systems
that continue to be available after the release of this book. Smaller companies are
often bought by bigger companies that operate in the same space, and that try to
increase their market share and provide novel and innovative products.

A further difficulty is that for most commercial products in this area not many
technical details are published that would allow an objective technical discussion
and comparison of their capabilities. Often only white papers and sales brochures
are available from a vendor of a commercial data matching system. Such white
papers do generally not contain any details about the algorithms or technologies
implemented in a product, which is understandable given most commercial systems
are based on some sort of proprietary technology developed by the companies that
sell the product.

P. Christen, *Data Matching*, Data-Centric Systems and Applications,
DOI: 10.1007/978-3-642-31164-2_10, © Springer-Verlag Berlin Heidelberg 2012

Herzog et al. [143] have provided a comprehensive checklist for evaluating commercial data matching software, which is based on an extensive earlier report and checklist by Day [90]. They key questions covered by these checklists are:

- What type of data matching can be conducted by a product? Deduplication of one file or database, and/or matching of two (or more) files or databases?
- Does the product include any functionality to pre-process (clean, standardise, and segment) data before matching?
- What indexing (blocking) techniques and what comparison functions are implemented in a product?
- What is the underlying matching technique or methodology used? Probabilistic record linkage, rule-based, or an advanced (machine learning) based approach? What parameters are available to the user, how well are they described, and are there mechanisms that support the tuning of these parameters to achieve an optimal matching outcome?
- Is the product specialised for a certain application domain or generally applicable? Special domains can include the deduplication of (business) mailing lists, the matching of health databases, or name matching for identity verification.
- Is there a limitation of the product in the size of the files or databases that can be handled, or is the limitation only based on the computing platform used?
- On what computing system can the product run, and what are the hardware requirements? Does the product make use of modern processors by exploiting multi-core parallelism or distributed systems?
- What types of data can be accessed? This can include comma or tabulator separated text files, fixed column width text files, spreadsheets, proprietary binary files, or database access through Open Database Connectivity (ODBC) or via a standard SQL interface.
- Does the product provide a graphical user interface, a command line interface, and/or an Application Programming Interface (API) that allows matching routines to be called from other programs?
- Are the licencing arrangements limiting what kind of data can be matched, and how many users can work with a product at any one time?
- How well is the product documented, what are the support arrangements by the supplier of the product, and what are the costs (both purchase upfront as well as annual licencing and support fees)?

A search for 'deduplication software' or 'data matching software' on any major Web search engine will provide a large number of links to products and services that offer deduplication or matching capabilities.

A valuable resource for practitioners working in the area of data matching and deduplication is the *Data Matching* group of the *LinkedIn* professional online network, which is accessible at: http://www.linkedin.com/groups/Data-Matching-2107798.

10.2 Research and Open Source Systems

The following sections provide an overview of freely available data matching systems that have mostly been developed by researchers as part of their work of inventing new and improved data matching algorithms and techniques. Some of these systems include graphical user interfaces and extensive documentation, while others are more basic research prototypes consisting of a set of program codes only. A comprehensive list to data matching software implementations is also available at: http://en.wikipedia.org/wiki/record_linkage.

The systems presented in the following sections are alphabetically ordered. Only systems for which implementations are available are covered. A recent survey by Köpcke and Rahm also provides a comparative evaluation of eleven data matching systems [168]. The comparison criteria in this survey include both the functionalities provided by the different systems with regard to the different steps of the data matching process, as well as their performance with regard to matching accuracy and complexity measured on several data sets from different domains.

10.2.1 BigMatch

The BigMatch system has been developed and is being used by the US Census Bureau to match very large census data collections [295]. BigMatch is not a full data matching system, rather it is a program that can be used to extract plausible (potential) matches from very large files that otherwise could not be processed. These plausible matches are saved into several smaller files so that they can be individually processed with a proper data matching system later on.

BigMatch assumes the matching of a very large file, called the 'Record' file, with a smaller file, called the 'Memory' file. As the name indicates, the second file will be fully loaded into the main memory of a large compute server using efficient index data structures. The main 'Record' file then only needs to be read and processed once. Each record in this larger file is compared with the records stored in main memory, and all plausible matches are saved into new smaller result files. The matching of the 'Record' file with the 'Memory' file is conducted using a standard blocking approach with several blocking criteria, as was discussed in Sects. 4.2 and 4.4. For each record in the large 'Record' file that has a blocking key value that also occurs in the smaller 'Memory' file, a match similarity value is calculated and saved into a result file.

The BigMatch program is written in C, and in 2007 it was reported that it can match around 300,000 record pairs per second when using ten blocking criteria [295]. The program does not include a graphical user interface, rather it is controlled using two text based configuration files. The first configuration file simply contains the paths and names of the input files, while the second configuration file contains the parameters that control the running of the program. These parameters include the number of blocking criteria, the number of attributes in the input files, flags for output cut-off values, and (if known) the number of records in the input files. For each blocking criteria, the parameter file then contains details about which attributes to use for a blocking pass. Next, the details about which attributes are to be compared and how (exact or approximate string comparison, numerical age or numerical year comparison) need to be provided in the parameter file. The input files themselves are assumed to be flat text files with fixed record length. Parallel versions of BigMatch have been developed by the US Census Bureau and they are in use for large production data matching projects. Further information about BigMatch is available from the developers [295].

10.2.2 D-Dupe

D-Dupe is a graphical tool developed by the Department of Computer Science at the University of Maryland that allows the interactive exploration of networks that contain duplicates [36, 156]. It combines data matching algorithms with network visualisations. Users can explore subnetworks that potentially contain duplicates, and the tool will highlight relationships in these subnetworks. Further duplicates might be revealed when a user merges a pair of nodes in a network (labels them as being a unique entity). This approach improves the efficiency of the interactive deduplication conducted by a user.

Figure 10.1 shows the main screen of the D-Dupe system. Both string and relational comparison functions are implemented. The former include the edit distance, Jaro, Jaccard and Monge-Elkan comparison functions described in Chap. 5. Relational similarities are calculated based on the overlap of their common neighbouring entities [156]. To narrow down the search of duplicates, an indexing technique based on standard blocking has been implemented in the D-Dupe tool. A user can also directly edit the actual attribute values in a data set.

The D-Dupe software (as Windows binary), related publications, and a video demonstration is available from: http://www.cs.umd.edu/projects/linqs/ddupe/.

D-Dupe is implemented in C#, and for approximate string comparison functions is uses the SimMetrics package described in Sect. 10.2.11 [156]. Licencing options for both commercial and non-commercial use are available.

10.2.3 DuDe

The 'Duplicate Detection' (DuDe) system is a toolkit developed at the University of Potsdam [95]. DuDe contains several modules each with a well-defined interface. The 'DataSource' module provides methods to access to comma separated values (CSV) files, XML documents, JSON files, and bibliographies in Bibtex format. The 'Preprocessor' module is optional, in that it only is required if statistics about the input data sets are to be gathered, such as the frequency distributions of attribute values

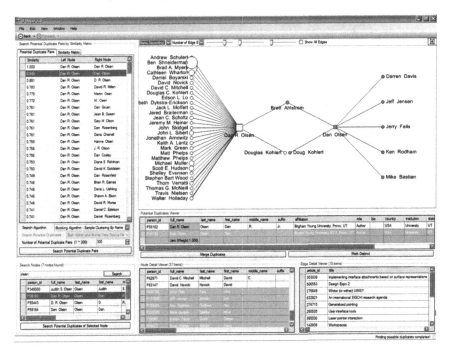

Fig. 10.1 A screenshot of the graphical user interface of the D-Dupe system which allows the interactive exploration of networks that can contain duplicates. The user can select various comparison functions, and manually merge two nodes if they correspond to the same entity. Potential duplicates are shown in the upper left area, while the network in the upper right area shows the relational contexts of the selected two records (in this example their co-author relationship). Details of the actual records are shown in the lower right area

needed to calculate for example the term-frequencies required for certain approximate string comparison functions. The main 'Algorithm' module then contains a series of indexing techniques that generate candidate record pairs from the input data sets. These pairs are then compared in the 'SimilarityFunction' module, which contains various types of comparison functions, as well as functions that aggregate similarity values. The 'Postprocessor' module allows calculation of the transitive closure, as well as gathering various statistics. Finally, the 'DuDeOutput' module provides functions to write the results of a deduplication or data matching project into different types of output formats, including CSV and JSON files.

The modular structure of the DuDe system allows users to develop their own code to replace or extend the functionality provided by DuDe, for example by developing a new indexing or comparison function. The DuDe system is written in Java and is available from: http://www.hpi.uni-potsdam.de/naumann/projekte/dude_duplicate_detection.html.

DuDe does not include a graphical user interface, rather it is configured using Java programs that define data access, and which algorithms to use for indexing, attribute

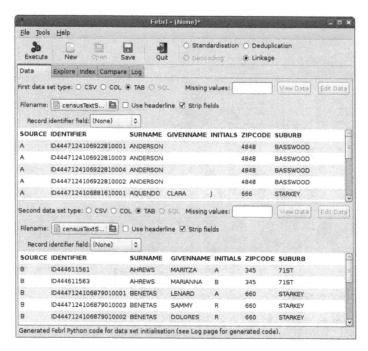

Fig. 10.2 A screenshot of the graphical user interface of the FEBRL data matching system, shown with two data sets initialised for a matching (Linkage) project. The data sets shown are based on the 'Census' data set described in Sect. 7.5

similarity comparisons, classification, and evaluation. The basic idea is for a user to write a Java program which uses the functionality of the modules provided by DuDe. Extensive user documentation, including example programs, are available on the DuDe Website given above. Three test data sets together with their gold standard (known matching results) are also included in the DuDe toolkit [95].

10.2.4 FEBRL

The 'Freely Extensible Biomedical Record Linkage' (FEBRL) system is an open source data matching system that has been developed since 2003 at the Australian National University as part of a collaborative research project with the New South Wales Department of Health [61, 62, 66]. The aims of this project were to develop techniques for improved data pre-processing, deduplication and data matching. While the focus of the project was on the application of these techniques to health databases, the FEBRL software itself is not limited to health data but is applicable to data from other domains as well. FEBRL is hosted on the *Sourceforge.Net* open source software repository, and it is available from: https://sourceforge.net/projects/febrl/.

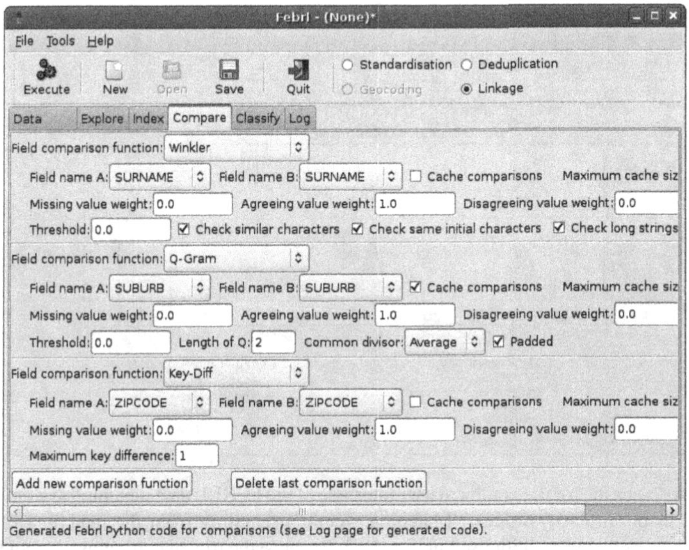

Fig. 10.3 A screenshot of the graphical user interface of the FEBRL data matching system, showing the definitions of three different comparison functions applied on the SURNAME, SUBURB and ZIPCODE attributes

FEBRL contains modules for data pre-processing of names, addresses, telephone numbers, and dates (based on rules and hidden Markov models as was discussed in Chap. 3); indexing (with seven indexing and eight phonetic encoding techniques implemented); comparisons (containing 26 comparison functions for strings, numbers, date, times and geographic locations); classification (with six classifiers); and evaluation. In its current version it allows access to text data sets in CSV, tabulator separated values (TAB), and fixed column width (COL) formats.

The software is written in the Python[1] programming language, which is an ideal language for rapid prototype development. Python can efficiently handle large data sets consisting of strings (which is the most common data type required for data matching), and it also provides modules for database access, parallel computing, Web programming, and for the development of graphical user interfaces (GUIs).

Users can inspect all source code modules of FEBRL, which allows them to learn about the different data matching techniques implemented in FEBRL. Due to its modular structure, FEBRL also allows new algorithms to be implemented and tested rapidly. FEBRL includes a GUI which allows non-expert users to experiment with different data matching techniques without the need to understand the details of the Python programming language. Figures 10.2, 10.3 and 10.4 show several screenshots of the FEBRL GUI.

FEBRL contains a selection of small test data sets, including the first four data sets described in Sect. 7.5. Additionally, FEBRL includes a data generator which

---

[1] http://www.python.org.

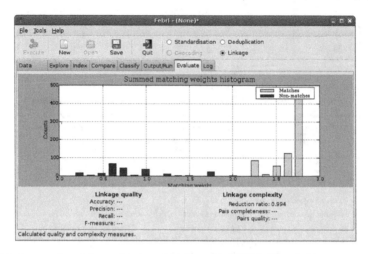

Fig. 10.4 A screenshot of the graphical user interface of the FEBRL data matching system, showing the histogram of similarity values ('Matching weight') resulting from the matching of the two 'Census' data sets that were initialised in Fig. 10.2. Because the true match status was not known in this example, only the complexity measure 'Reduction ratio' (Eq. 7.7 on p. 174) could be calculated

allows the creating of data sets according to a large number of parameters that can be set by the user [56, 72], as was previously discussed in Sect. 7.6.

10.2.5 FRIL

The 'Fine-Grained Records Integration and Linkage' (FRIL) tool was developed as part of a collaboration between Emory University and the Centers for Disease Control and Prevention (CDC) [154]. FRIL contains an extensive collection of parameters that can be set by the user, allows schema reconciliation as well as data matching, contains several indexing methods (including standard blocking described in Sect. 4.4 and the sorted neighbourhood approach described in Sect. 4.5), a variety of comparison functions, and a parameter tuning approach based on the expectation maximisation (EM) algorithm. FRIL can be run on multi-core systems and it contains a GUI that allows users to easily set-up and customise deduplication or data matching projects. Figures 10.5, 10.6 and 10.7 show different screenshots of the FRIL system. FRIL is written in Java and available for download from: http://fril.sourceforge.net/.

FRIL allows a user to select parameters such as matching fields, comparison functions, matching weights, indexing functions, and which evaluation metrics to employ. It also allows the pre-processing of attributes through the use of regular expressions to standardise the input data, and to split and merge attributes before they are used for matching. Multi-core parallelisation is transparent to the user,

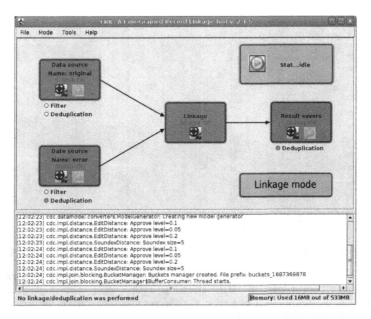

Fig. 10.5 A screenshot of the main FRIL window configured for the linkage of two data sets. The icons shown allow the configuration of different aspects of the data matching process

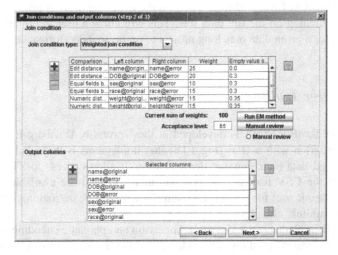

Fig. 10.6 A screenshot of the FRIL window that allows the configuration of the matching (join) conditions to be specified. As can be seen, different comparison functions are available, and weights can be assigned to individual attribute comparisons. The expectation–maximisation (EM) algorithm can be employed to tune parameter values. Also shown is the configuration of how the matched data are to be written into an output file

#	Confidence	name@original / n...	DOB@original / DO...	sex@original / sex...	race@original / ra...	weight@original / ...	height@original / ...
1	79	Aaron HARDING	112258	0			69
		Aaron HARDING	112258	0	3		69
2	90	Javon SHAFER	101042	0	4		79
		Javon SHAFER	101042	0	4	142	79
3	80	Alivia NUNEZ	012349	0	4	111	
		Alivia NUNEZ	012349	0	4		64
4	70	Karen SALMON	080744	0	4	164	69
		Karen SALMON	080744	0	4	145	56
5	90	Cecilia MASTERS	080965	0	1		72
		Cecilia MASTERS	080965	0	1	156	72
6	72	Devon LACKEY	051078	0	1	154	74
		Devon LACKEY	051078	0	1	167	85
7	75	Paris CROSBY	012375	0	1	107	55
		Paris CROSBY	012375	0	1		44
8	86	Javon CALL	040867	0	3	151	
		Javon CALL	040867	0	3	147	
9	75	Hayley IRVIN	022675	0	2	124	69
		Hayley IRVIN	022675	0	2	99	
10	80	Julie BONE	020483	0	4		
		Julie BONE	020483	0	4	155	77
11	85	Sierra ROMANO	092359	0	3	97	60
		Sierra ROMANO	092359	0	3	112	60
12	75	Cecilia MARINO	060563	0	3	115	
		Cecilia MARINO	060563	0	3	101	60
13	83	Pierce GRAY	042177	0	2	132	
		Pierce GRAY	042177		2	132	61
14	75	Alec LEONE	062248	0	1	166	
		Alec LEONE	062248	0	1	199	85

200 linkages loaded. Page 1 of 15.   Filter: O | Sort: O ◄ 1 ►

Fig. 10.7 A screenshot of the FRIL clerical review window, where pairs of records and their attribute values are shown. Cells in *green* indicate the same attribute values while *red* cells highlight different values. Each record pair is also given a value of confidence that corresponds to the likelihood that the two records refer to the same entity

which means that FRIL will detect if it is run on a multi-core system and therefore run its algorithms in parallel. An 'Analysis Window' allows the inspection of the parameters used by FRIL and their effect on the matching process prior to running the actual matching on the full data sets. This will allow a user to efficiently tune parameters for a given data matching project to optimise its performance.

10.2.6 Merge ToolBox

The 'Merge ToolBox' (MTB) developed at the University Duisburg-Essen [238] is a Java program with a GUI (Fig. 10.8) that implements both probabilistic and distance-based data matching and deduplication. It allows access to STATA[2] and text files in the CSV format. The program and a manual are available (for free only for academic use) for download at: http://www.uni-due.de/soziologie/schnell_forschung_safelink_mtb.php.

The MTB implements a variety of comparison and phonetic encoding functions (including a German variation of Soundex called the 'Kölner Phonetik'), as well as the standard blocking technique described in Sect. 4.4. For probabilistic data matching, the match and non-match parameters (described in Sect. 6.3) can either be set manually or by using the EM algorithm.

A unique feature of the MTB is its ability to incorporate a privacy-preserving data matching approach based on hash-encoded Bloom filters [239]. The implementation

---

[2] www.stata.com.

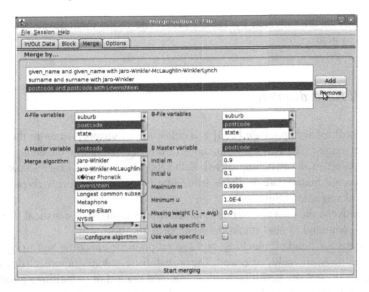

Fig. 10.8 A screenshot of the 'Merge ToolBox' user interface showing the configuration of comparison functions to be used on three different input attributes. As can be seen, approximate string comparison functions as well as phonetic encoding functions can be used to compare attribute values, and match (m) and non-match (u) parameters can be set to be used in the probabilistic matching approach (as was described in Sect. 6.3) implemented in the 'Merge ToolBox'

of this approach, called 'Safelink Prototype Software', is available from: http://www.
uni-due.de/soziologie/schnell_forschung_safelink_software.php.

Both Java and Python implementations of this prototype are available.

10.2.7 OYSTER

The 'Open sYSTem Entity Resolution' (OYSTER) system has been developed by the University of Arkansas, originally designed to provide entity resolution for student records and to allow longitudinal studies [249]. OYSTER can be run in several modes to conduct operations such as data matching/merge-purge, identity capture, and identity resolution. It contains modules for probabilistic matching, finding the transitive closure of matched record sets, and indexing. No internal database is used and all index data structures are kept in main memory. OYSTER is available under an open source software licence from: http://sourceforge.net/projects/oysterer/.

OYSTER is written in Java and uses XML sripts customised by the user to define entity attributes, the table layout of the data sources, and identity rules for resolving reference sources. The core of the OYSTER system is based on the R-Swoosh algorithm to entity resolution described in Sect. 6.12, but with an identity management system added. A detailed description of OYSTER, including example scripts

and screenshots are provided by Talburt [249], one of the main developers of the
OYSTER system.

10.2.8 R RecordLinkage

The 'RecordLinkage' package [233] is a collection of scripts written in the R statistical language.[3] The two main functions of this package allow the deduplication
of one data set or the matching of two data sets. The package, example data sets,
and a reference manual are available from: http://cran.r-project.org/web/packages/
RecordLinkage/index.html.

Alternatively, the package can be installed from within the R statistical language
using the command: install.packages('RecordLinkage').

Functions for standard blocking, and several phonetic encoding and string comparison methods, are included as well. The probabilistic matching approach described
in Sect. 6.3 is implemented together with the EM algorithm to facilitate parameter
estimation. Both supervised and unsupervised classification techniques (as available
through R packages) can be employed, and an edit function facilitates manual clerical
review of potential matches, as was discussed in Sect. 7.4.

10.2.9 SecondString

The SecondString toolkit is a set of Java classes that implement many approximate string comparison functions [84], including most of the comparisons functions
described in Chap. 5. The software and documentation can be downloaded from:
http://sourceforge.net/projects/secondstring/.

Besides basic string comparison functions, SecondString also contains implementations of the SoftTFIDF and the Monge-Elkan multi-word comparisons functions
(as described in Sects. 5.8 and 5.6), and a combined adaptive string distance learner
which can be trained using example pairs of strings that are known to either correspond to true matches or true non-matches.

10.2.10 SILK

SILK is a system developed at the Free University of Berlin that allows the discovery
of relationships in data within the Linked Data framework [150, 264]. These data are
assumed to be represented as RDF (Resource Description Framework) tuples. This

---

[3] http://www.r-project.org/.

means that data publishers can employ SILK to match their RDF data with data from other sources. SILK is available at: http://www4.wiwiss.fu-berlin.de/bizer/silk/.

SILK is implemented in Java and contains indexing and approximate string comparison functions, and it facilitates parallel execution using the Map-Reduce framework. The string comparison module of the FEBRL data matching system (presented in Sect. 10.2.4) is used for the approximate string comparison functionality in SILK. Matching conditions can be specified by a user using a declarative link specification language, detailing what conditions must be fulfilled in order for RDF tuples of different types to be classified as matches.

10.2.11 SimMetrics

Similar to SecondString, SimMetrics is a Java package containing a large number of approximate string comparison functions. It is available on *Sourceforge.Net* at: http://sourceforge.net/projects/simmetrics/.

While the original Website of SimMetrics is no longer available, both the Java source code as well as its documentation are available at the above URL.

Besides string comparison functions that are commonly used for data matching, SimMetrics also includes several functions that are more suitable for comparing long sequences (like genome sequences), such as the Needleman–Wunsch, Smith–Waterman–Gotoh, and the Smith–Waterman algorithms, the latter of which was previously described in Sect. 5.3.1.

10.2.12 TAILOR

The 'RecOrd LinkAge Toolbox' (TAILOR) was developed at Purdue University in the early 2000 [102] as an extensible toolbox that allows different indexing, comparison, and classification techniques, as well as various evaluation methods, to be integrated. Standard blocking and the sorted neighbourhood approach are implemented in TAILOR, as are several comparison and phonetic encoding functions. With regard to classification techniques, TAILOR contains both unsupervised (clustering) and supervised (decision tree) techniques, as was described in Sects. 6.6 and 6.9. The user can interact with TAILOR either through a definition language or via a GUI. TAILOR is written in Java and is available by contacting the developers [102].

10.2.13 WHIRL

The 'Word-Based Heterogeneous Information Representation Language' (WHIRL) system [81, 82] was developed by William Cohen in the late 1990s. WHIRL can be seen as a 'soft' database management system that allows various similarity comparison functions (similarity joins) to be applied on textual data, which are represented using the vector space model that was discussed in Sect. 5.8. The WHIRL system is written in C++ and available from: http://www.cs.cmu.edu/~wcohen/whirl/.

The data model used in WHIRL, called 'Storing Texts In Relations' (STIR), together with a query language that allows conjunctive queries, is an approach similar to the rule-based classification approach that was discussed in Sect. 6.5. The results of a 'matching' in WHIRL is a ranked list of answers (database records or tuples) in a similar way as Web search engines return a ranked list of search results relevant to a query.

Glossary

The following glossary draws from earlier glossaries on data matching that were provided by Newcombe (1988) [199], Day (1995) [90], Gill (2001) [119], Blakely and Salmond (2002) [37], and Talburt (2011) [249]. The description of terms follows these earlier glossaries, however, various modifications have been made to improve the consistency of the terminology used. New terms that are not listed in previous glossaries, but that are relevant to the topics covered in this book, have been added. Any occurrence of a term that is also described in the glossary is shown in *italics*.

Active learning A type of *classifier* that builds a *classification model* based on the similarity values in *comparison vectors* and optionally the relationships between pairs or groups of *records*. In order to improve its classification model, an active learning classifier asks for manual feedback on the *match status* of selected *candidate record pairs* (that are difficult to classify). An iterative process of manual feedback and building an improved classification model is carried out until a certain stopping criterion has been achieved.

Agreement weight See *match weight*.

Approximate match The status of a *candidate record pair* that has been compared using an *approximate matching* process, and where one or several *attribute values* that have been compared are different between the two records.

Approximate matching The process by which *candidate record pairs* are compared using a set of *comparison functions* that allows for approximate (not exact) similarities.

Attribute A column in a database table, file, or spreadsheet, that contains a well-defined type of data, such as strings, numbers, dates, times and so on.

Attribute value A value stored in a specific *attribute* and a specific *record* in a database, file or spreadsheet.

P. Christen, *Data Matching*, Data-Centric Systems and Applications,
DOI: 10.1007/978-3-642-31164-2, © Springer-Verlag Berlin Heidelberg 2012

Blocking A type of *indexing* technique that has traditional been employed in *data matching* to reduce the number of *record pairs* that need to be compared. Blocking splits the input database(s) according to a *blocking key*. Only *records* that have the same *blocking key value* are inserted into the same block. *Candidate record pairs* are formed from all records in the same block.

Blocking key Also called 'blocking variable', a blocking key defines how one or more *attribute values* from *records* in the input database(s) are processed (often using *phonetic encoding* algorithms) and concatenated into *blocking key values* during the *indexing* step of the *data matching* process. A good blocking key should lead to records that are similar with each other to be inserted into the same block, and records that are dissimilar to each other to be inserted into different blocks.

Blocking key value A string value for an individual *record* in an input database which has been generated using a *blocking key* definition. The blocking key value for a record determines into which block(s) the record is being inserted during the *indexing* step of the *data matching* process.

Candidate record pair A *record pair* that is formed from the *records* that have been inserted into the same block (or cluster or window) by an *indexing* technique. All candidate record pairs are compared in detail in the comparison step of the *data matching* process using various *comparison functions*.

Classification model A model that determines how *candidate record pairs* are classified into *matches, non-matches* and optionally *potential matches*. A classification model is generated by a *classifier* algorithm.

Classifier A type of algorithm that builds a *classification model* based on an *supervised learning, unsupervised learning* or *active learning* approach.

Cleaning The process of removing unwanted characters and tokens (alphanumerical words) from the *attribute values* in the input database(s) during the *data pre-processing* step of the *data matching* process.

Clerical review The process of manually assessing and classifying as *matches* or *non-matches* the *candidate record pairs* that have been classified as *potential matches* in a *classifier* such as *probabilistic record linkage* or a related approach.

Clustering A type of algorithm that groups similar data objects (*records* in the case of *data matching*) together according to the similarities calculated between these objects. In *data matching* and *deduplication*, the similarities between records are captured in the *comparison vector* for each *candidate record pair*. The objectives of clustering in data matching and *deduplication* are to (1) have each generated cluster to correspond to one *entity*, and (2) each entity stored in the database(s) that are matched or deduplicated is assumed to be represented by one cluster only.

Collective entity resolution A type of algorithm that aims to solve the *entity resolution* problem in an overall collective fashion by considering the relationships between the *records* stored in the database(s) to be matched or deduplicated as a graph. Based on such a relationship graph, a collective entity resolution approach classifies groups of records into *matches* and *non-matches*. This approach is in contrast to the *pair-wise classification* of individual record pairs.

Comparison function A function that has as input two *attribute values* (which can be strings, numbers, dates, times or more complex objects) and that calculates a similarity between these two values. The comparison can either be exact or allow for approximate similarity. An exact comparison function generally returns a similarity value of 0 if the two attribute values are different from each other, or a similarity value of 1 if they are the same. An approximate comparison function generally returns a normalised numerical similarity value between 0 and 1 that indicates the similarity between the two attribute values, with a larger similarity value indicating a higher similarity between the two attribute values. If only exact comparison functions are used when attribute values are compared then this process corresponds to *exact matching*, while when approximate comparison functions are used the process corresponds to *approximate matching*. A popular class of approximate comparison functions are approximate *string comparison functions*.

Comparison vector The vector of similarity values generated for a *candidate record pair* when one or more *attributes* of the pair are compared using *comparison functions* that are appropriate to the content of the attributes. If *n* comparison functions are used then the resulting comparison vector will contain *n* similarity values.

Data linkage The name used by statisticians, and health and biomedical researchers and practitioners, for the process of *data matching*.

Data matching The process of comparing *records* from two or more databases with the objective to identify pairs or groups of records that refer to the same *entity*. These pairs or groups of records are known as *matches*.

Data pre-processing The process of *cleaning, standardising*, and *segmenting*, the *attribute values* stored in the input database(s) to be matched or deduplicated, with the general aim to improve data quality, and more specifically, to improve the outcomes of the *data matching* or *deduplication* process.

Deduplication The process of *duplicate detection* followed by a process which for each *entity* in a database either merges the identified *duplicate* records into one combined record, or removes some records from the database until it only contains a single record for each entity.

Disagreement weight See *non-match weight*.

Duplicates The presence in a single database of multiple *records* that refer to the same *entity*.

Duplicate detection The process of comparing *records* from a single database with the objective to identify pairs or groups of records that refer to the same *entity*. These pairs or groups of records are known as *duplicates*.

Entity A real-world subject or object, such as an individual person, business, publication or consumer product, that has a unique identity and that can be distinguished from any other entity.

Entity identifier A number, code or string that uniquely identifies a single *entity* within the database(s) that are matched or deduplicated.

Entity resolution The process of comparing *records* from one or more databases with the objective to identify pairs or groups of records that refer to the same *entity*, to classify these pairs or groups as *matches* (and pairs or groups of records that do not refer to the same entity as *non-matches*), and to merge all records that refer to the same entity into a new combined record. The result of an entity resolution process is a set of combined records that each corresponds to one entity, and each of the entities stored in the database(s) that were matched is represented by a single combined record only. Entity resolution applied on a single database is also known as *deduplication*.

Exact match The status of a *candidate record pair* that has been compared using either an *exact matching* or an *approximate matching* process, and where all *attribute values* that have been compared are the same in both records of the pair.

Exact matching The process by which *candidate record pairs* are compared using a set of *comparison functions* that only permit exact similarities.

False match A *record pair* that is classified as a *match* where, however, the two *records* of the pair refer to two different *entities*. In the context of classification a false match is also known as a false positive.

False non-match A *record pair* that is classified as a *non-match* where, however, both *records* in the pair correspond to the same *entity*. In the context of classification a false non-match is also known as a false negative.

Indexing The process of splitting a database into smaller blocks or clusters, or sorting a database, with the aim to reduce the number of *record pair* comparisons that are conducted. *Records* that have the same *blocking key value* are inserted into the same block or cluster, or they are sorted close to each other if they have the same or a very similar *sorting key value*. *Candidate record pairs* are formed from all records that are in the same block or cluster, or that appear in the same window. The traditional approach used for indexing has been *blocking*.

Information extraction The process of identifying and extracting instances of particular classes of entities, events, or relationships, from unstructured data such as natural text, and their transformation into a structured representation such as a database *record*.

Match A pair or group of *records* that is classified as referring to the same *entity*.

Match status The outcome of applying a *classification model* on a *candidate record pair*. This outcome can be that the record pair is classified as a *match*, a *non-match*, and (optionally) as a *potential match*. For a *classifier* such as *probabilistic record linkage* or a related approach, the potential matches are the candidate record pairs that are not clear matches or non-matches. These pairs need to be manually assessed and classified in a *clerical review* process.

Match weight In *probabilistic record linkage* and related *classifier* approaches, a match weight is a numerical value that is assigned to a certain *attribute* where the *attribute values* are the same or similar to each other (assumed to be in agreement). A match weight is also called an *agreement weight*. Match weights are calculated as the likelihood that two attribute values are in agreement assuming that both records in a *candidate record pair* correspond to the same entity, divided by the likelihood that two attribute values are in agreement assuming that the two records in a candidate record pair correspond to different entities.

Merge/purge The name used by database and data warehousing researchers and practitioners for the process of *data matching* and *deduplication*.

Non-match A pair or group of *records* that is classified as referring to different entities.

Non-match weight In *probabilistic record linkage* and related *classifier* approaches, a non-match weight is a numerical value that is assigned to a certain *attribute* where the *attribute values* are different from each other (not assumed to be in agreement). A non-match weight is also called a *disagreement weight*. Non-match weights are calculated as the likelihood that two attribute values are in disagreement assuming that both records in a *candidate record pair* correspond to the same entity, divided by the likelihood that two attribute values are in disagreement assuming that the two records in a candidate record pair correspond to different entities.

Pair-wise classification A type of algorithm that classifies individual *candidate record pairs* into *matches*, *non-matches*, and (optionally) into *potential matches*, without taking the *match status* of other candidate record pairs into account. This classification approach is in contrast to *collective entity resolution* techniques that aim to classify all candidate record pairs collectively in an overall optimal fashion.

Parsing See *segmenting*.

Phonetic encoding A type of algorithm that converts a string (generally assumed to correspond to a name) into a code that represents the pronunciation of that string. Popular phonetic encoding algorithms include Soundex, NYSIIS, ONCA, Phonex, Phonix and Double-Metaphone.

Potential match A *candidate record pair* that is classified in *probabilistic record linkage* or a related *classifier* approach to *data matching* as potentially referring to the same *entity*. The final *match status* for these candidate record pairs is determined through a manual assessment in a *clerical review* process.

Privacy-preserving data matching Also known as privacy-preserving record linkage, this is the process of matching databases from different organisations such that none of the database owners has to reveal any of their private or confidential data, and at the end of the matching process only limited information, such as the number of *records* that have been classified as *matches*, or only their *record identifiers*, is being revealed to the database owners.

Probabilistic record linkage A statistical *classifier* approach to *data matching* published by Fellegi and Sunter in 1969 [108]. This approach calculates *match weights* and *non-match weights* based on error probabilities and frequency distributions of *attribute values* in the input databases. *Candidate record pairs* are classified based on their *weight vectors* into either *matches*, *non-matches*, or *potential matches*, using a threshold-based and *pair-wise classification* approach.

Record A row in a database table, file, or spreadsheet, that contains values in a set of *attributes*. It is assumed that each record represents one *entity*, but that an entity can be represented by more than one record in a database, file or spreadsheet.

Record identifier A number, code, or string, that uniquely identifies a single *record* in a database. A record identifier is different from an *entity identifier*.

Record linkage See *data linkage*.

Record pair Two *records*, for the process of *data matching* one record from each of the two input databases that are matched, while for the *deduplication* of one database both records are sourced from the single input database.

Searching An alternative name for the *indexing* step of the *data matching* process. It is a term that is sometimes used for the process that is concerned with the reduction of the number of *candidate record pairs* that are generated. The aim of searching is to find candidate record pairs that likely refer to *matches*. Examples of searching techniques include *blocking*, sorting and hashing.

Segmenting The process of separating the tokens (white-space separated alphanumeric words) contained in an *attribute value* in the input database(s) into well-defined elements during the *data pre-processing* step of the *data matching* process. Segmenting is commonly based on some type of *information extraction* technique.

Similarity function See *comparison function*.

Sorting key Similar to a *blocking key*, a sorting key defines how the input database(s) are sorted for the sorted neighbourhood *indexing* technique [64, 140, 141]. A sorting key determines how values from certain *attributes* from the *records* in the input database(s) will be used to sort the databases. The aim of a sorting key is for similar records to be located closely to each other in the sorted database(s).

Standardisation The process of converting the *attribute values* in an input database into a standard format during the *data pre-processing* step of the *data matching* process, by, for example, converting all letters into lower or upper case, by correcting misspellings and replacing nicknames with their proper names, and by expanding abbreviations into full words.

String comparison function A type of *comparison function* that takes as input two strings and that returns an exact similarity value (*exact matching*) or an approximate similarity value (*approximate matching*) calculated for the two input strings.

Supervised learning A type of *classifier* algorithm that builds a *classification model* based on the similarity values in *comparison vectors* and optionally the relationships between pairs or groups of *records*. A supervised classification model is built based on training data that are in the form of pairs or groups of *records* where their *match status* (*true match* and *true non-match*) is known.

Transitive closure The process of deciding the *match status* for all *candidate record pairs* in a group of *records* as *matches*, where some but not all individual pairs of records have been classified as matches, following the transitivity property of the match classification [195].

True match A *record pair* that is classified as a *match*, where both *records* in the pair correspond to the same *entity*. In the context of classification a true match is also known as a true positive.

True non-match A *record pair* that is classified as a *non-match*, where the two *records* in the pair correspond to two different *entities*. In the context of classification a true non-match is also known as a true negative.

Weight vector A vector containing numerical values for a *candidate record pair*. Weight vectors are used by *probabilistic record linkage* and related *classifier* approaches to decide the *match status* of candidate record pairs. The values in the weight vector of a candidate record pair are calculated by combining for each compared *attribute* the *match weight* (if *attribute values* are the same or similar) or *non-match weight* (if attribute values are different) for that attribute with the similarity value for that attribute taken from the pair's *comparison vector*.

Unsupervised learning A type of *classifier* algorithm that builds a *classification model* based on the similarity values in *comparison vectors* and optionally the relationships between pairs or groups of *records*. The classification model is built without knowing the true *match status* of these pairs or groups of records. A popular type of unsupervised learning algorithm is *clustering*.

References

1. Adly, N.: Efficient record linkage using a double embedding scheme. In: DMIN, pp. 274–281. Las Vegas (2009)
2. Aggarwal, C.C.: Managing and Mining Uncertain Data, *Advances in Database Systems*, vol. 35. Springer (2009)
3. Aggarwal, C.C., Yu, P.S.: The IGrid index: Reversing the dimensionality curse for similarity indexing in high dimensional space. In: ACM SIGKDD, pp. 119–129. Boston (2000)
4. Aggarwal, C.C., Yu, P.S.: Privacy-preserving data mining: models and algorithms, *Advances in Database Systems*, vol. 34. Springer (2008)
5. Agichtein, E., Ganti, V.: Mining reference tables for automatic text segmentation. In: ACM SIGKDD, pp. 20–29. Seattle (2004)
6. Agrawal, R., Evfimievski, A., Srikant, R.: Information sharing across private databases. In: ACM SIGMOD, pp. 86–97. San Diego (2003)
7. Aizawa, A., Oyama, K.: A fast linkage detection scheme for multi-source information integration. In: WIRI, pp. 30–39. Tokyo (2005)
8. Al-Lawati, A., Lee, D., McDaniel, P.: Blocking-aware private record linkage. In: International Workshop on Information Quality in Information Systems, pp. 59–68 (2005)
9. Alvarez, R., Jonas, J., Winkler, W., Wright, R.: Interstate voter registration database matching: the Oregon-Washington 2008 pilot project. In: Workshop on Trustworthy Elections, pp. 17–17. USENIX Association (2009)
10. Anderson, K., Durbin, E., Salinger, M.: Identity theft. The Journal of Economic Perspectives **22**(2), 171–192 (2008)
11. Arasu, A., Götz, M., Kaushik, R.: On active learning of record matching packages. In: ACM SIGMOD, pp. 783–794. Indianapolis (2010)
12. Arasu, A., Kaushik, R.: A grammar-based entity representation framework for data cleaning. In: ACM SIGMOD, pp. 233–244. Providence, Rhode Island (2009)
13. Armstrong, M.P., Ruggles, A.J.: Geographic information technologies and personal privacy. Cartographica: The International Journal for Geographic Information and Geovisualization **40**(4), 63–73 (2005)
14. Atallah, M., Kerschbaum, F., Du, W.: Secure and private sequence comparisons. In: Workshop on Privacy in the Electronic Society, pp. 39–44. ACM (2003)
15. Baeza-Yates, R., Ribeiro-Neto, B.: Modern Information Retrieval. Addison-Wesley Longman Publishing Co., Boston (1999)
16. Baldwin, J., Acheson, E., Graham, W.: Textbook of medical record linkage. Oxford University Press (1987)

17. Barone, D., Maurino, A., Stella, F., Batini, C.: A privacy-preserving framework for accuracy and completeness quality assessment. Emerging Paradigms in Informatics, Systems and Communication p. 83 (2009)

18. Bartolini, I., Ciaccia, P., Patella, M.: String matching with metric trees using an approximate distance. In: String Processing and Information Retrieval, LNCS 2476, pp. 271–283. Lisbon, Portugal (2002)

19. Batini, C., Scannapieco, M.: Data quality: Concepts, methodologies and techniques. Data-Centric Systems and Applications. Springer (2006)

20. Baxter, R., Christen, P., Churches, T.: A comparison of fast blocking methods for record linkage. In: ACM SIGKDD Workshop on Data Cleaning, Record Linkage and Object Consolidation, pp. 25–27. Washington DC (2003)

21. Bayardo, R., Ma, Y., Srikant, R.: Scaling up all pairs similarity search. In: WWW, pp. 131–140. Banff, Canada (2007)

22. Behm, A., Ji, S., Li, C., Lu, J.: Space-constrained gram-based indexing for efficient approximate string search. In: IEEE ICDE, pp. 604–615. Shanghai (2009)

23. Belin, T., Rubin, D.: A method for calibrating false-match rates in record linkage. Journal of the American Statistical Association pp. 694–707 (1995)

24. Bellahsene, Z., Bonifati, A., Rahm, E.: Schema Matching and Mapping. Data-Centric Systems and Applications. Springer (2011)

25. Benjelloun, O., Garcia-Molina, H., Gong, H., Kawai, H., Larson, T., Menestrina, D., Thavisomboon, S.: D-Swoosh: A family of algorithms for generic, distributed entity resolution. In: International Conference on Distributed Computing Systems, pp. 37–37 (2007)

26. Benjelloun, O., Garcia-Molina, H., Menestrina, D., Su, Q., Whang, S., Widom, J.: Swoosh: a generic approach to entity resolution. The VLDB Journal 18(1), 255–276 (2009)

27. Bergroth, L., Hakonen, H., Raita, T.: A survey of longest common subsequence algorithms. In: String Processing and Information Retrieval, pp. 39–48. A Curuna, Spain (2000)

28. Bernecker, T., Kriegel, H.P., Mamoulis, N., Renz, M., Zuefle, A.: Scalable probabilistic similarity ranking in uncertain databases. IEEE Transactions on Knowledge and Data Engineering 22(9), 1234–1246 (2010)

29. Bertolazzi P De Santis L, S.M.: Automated record matching in cooperative information systems. In: Proceedings of the international workshop on data quality in cooperative information systems. Siena, Italy (2003)

30. Bertsekas, D.P.: Auction algorithms for network flow problems: A tutorial introduction. Computational Optimization and Applications 1, 7–66 (1992)

31. Bhattacharya, I., Getoor, L.: Collective entity resolution in relational data. ACM Transactions on Knowledge Discovery from Data 1(1) (2007)

32. Bhattacharya, I., Getoor, L.: Query-time entity resolution. Journal of Artificial Intelligence Research 30, 621–657 (2007)

33. Bilenko, M., Basu, S., Sahami, M.: Adaptive product normalization: Using online learning for record linkage in comparison shopping. In: IEEE ICDM, pp. 58–65. Houston (2005)

34. Bilenko, M., Kamath, B., Mooney, R.J.: Adaptive blocking: Learning to scale up record linkage. In: IEEE ICDM, pp. 87–96. Hong Kong (2006)

35. Bilenko, M., Mooney, R.J.: Adaptive duplicate detection using learnable string similarity measures. In: ACM SIGKDD, pp. 39–48. Washington DC (2003)

36. Bilgic, M., Licamele, L., Getoor, L., Shneiderman, B.: D-dupe: An interactive tool for entity resolution in social networks. In: IEEE Symposium on Visual Analytics, Science and Technology, pp. 43–50 (2006)

37. Blakely, T., Salmond, C.: Probabilistic record linkage and a method to calculate the positive predictive value. International Journal of Epidemiology 31:6, 1246–1252 (2002)

38. Bleiholder, J., Naumann, F.: Data fusion. ACM Computing Surveys 41(1), 1–41 (2008)

39. Bloom, B.: Space/time trade-offs in hash coding with allowable errors. Communications of the ACM 13(7), 422–426 (1970)

40. Borgman, C.L., Siegfried, S.L.: Getty's synoname^{TM} and its cousins: A survey of applications of personal name-matching algorithms. Journal of the American Society for Information Science **43**(7), 459–476 (1992)
41. Borkar, V., Deshmukh, K., Sarawagi, S.: Automatic segmentation of text into structured records. ACM SIGMOD Record **30**(2), 175–186 (2001)
42. Breiman, L., Freidman, J., Olshen, R., Stone, C.: Classification and regression trees. Chapman and Hall/CRC (1984)
43. Broder, A., Carmel, D., Herscovici, M., Soffer, A., Zien, J.: Efficient query evaluation using a two-level retrieval process. In: ACM CIKM, pp. 426–434. New Orleans (2003)
44. Brook, E., Rosman, D., Holman, C.: Public good through data linkage: measuring research outputs from the Western Australian data linkage system. Australian and New Zealand journal of public health **32**(1), 19–23 (2008)
45. Brownstein, J.S., Cassa, C., Kohane, I.S., Mandl, K.D.: Reverse geocoding: Concerns about patient confidentiality in the display of geospatial health data. In: AMIA Annual Symposium Proceedings, p. 905. American Medical Informatics Association (2005)
46. Brownstein, J.S., Cassa, C., Mandl, K.D.: No place to hide-reverse identification of patients from published maps. New England Journal of Medicine **355**(16), 1741–1742 (2006)
47. Campbell, K., Deck, D., Krupski, A.: Record linkage software in the public domain: a comparison of Link Plus, The Link King, and a basic deterministic algorithm. Health Informatics Journal **14**(1), 5 (2008)
48. Cayo, M.R., Talbot, T.O.: Positional error in automated geocoding of residential addresses. International Journal of Health Geographics **2**(10) (2003)
49. Cebrián, M., Alfonseca, M., Ortega, A.: Common pitfalls using the normalized compression distance: What to watch out for in a compressor. Communications in Information and Systems **5**(4), 367–384 (2005)
50. Chambers, R.: Regression analysis of probability-linked data. Official Statistics Research Series **4** (2008)
51. Chan, Y., Talburt, J., Talley, T.: Data Engineering. Springer (2010)
52. Chaudhuri, S., Ganti, V., Motwani, R.: Robust identification of fuzzy duplicates. In: IEEE ICDE, pp. 865–876. Tokyo (2005)
53. Chaytor, R., Brown, E., Wareham, T.: Privacy advisors for personal information management. In: SIGIR Workshop on Personal Information Management, pp. 28–31. Seattle, Washington (2006)
54. Chen, H., Chung, W., Xu, J., Wang, G., Qin, Y., Chau, M.: Crime data mining: a general framework and some examples. IEEE Computer **37**(4), 50–56 (2004)
55. Chor, B., Kushilevitz, E., Goldreich, O., Sudan, M.: Private information retrieval. Journal of the ACM (JACM) **45**(6), 965–981 (1998)
56. Christen, P.: Probabilistic data generation for deduplication and data linkage. In: IDEAL, Springer LNCS, vol. 3578, pp. 109–116. Brisbane (2005)
57. Christen, P.: A comparison of personal name matching: Techniques and practical issues. In: Workshop on Mining Complex Data, held at IEEE ICDM. Hong Kong (2006)
58. Christen, P.: Privacy-preserving data linkage and geocoding: Current approaches and research directions. In: Workshop on Privacy Aspects of Data Mining, held at IEEE ICDM. Hong Kong (2006)
59. Christen, P.: Automatic record linkage using seeded nearest neighbour and support vector machine classification. In: ACM SIGKDD, pp. 151–159. Las Vegas (2008)
60. Christen, P.: Automatic training example selection for scalable unsupervised record linkage. In: PAKDD, Springer LNAI, vol. 5012, pp. 511–518. Osaka (2008)
61. Christen, P.: Febrl: An open source data cleaning, deduplication and record linkage system with a graphical user interface. In: ACM SIGKDD, pp. 1065–1068. Las Vegas (2008)
62. Christen, P.: Development and user experiences of an open source data cleaning, deduplication and record linkage system. SIGKDD Explorations **11**(1), 39–48 (2009)

63. Christen, P.: Geocode matching and privacy preservation. In: Workshop on Privacy, Security, and Trust in KDD, pp. 7–24. Springer (2009)
64. Christen, P.: A survey of indexing techniques for scalable record linkage and deduplication. IEEE Transactions on Knowledge and Data Engineering X(Y) (2011)
65. Christen, P., Belacic, D.: Automated probabilistic address standardisation and verification. In: AusDM, pp. 53–67. Sydney (2005)
66. Christen, P., Churches, T., Hegland, M.: Febrl—A parallel open source data linkage system. In: PAKDD, Springer LNAI, vol. 3056, pp. 638–647. Sydney (2004)
67. Christen, P., Churches, T., Willmore, A.: A probabilistic geocoding system based on a national address file. In: AusDM. Cairns (2004)
68. Christen, P., Churches, T., Zhu, J.: Probabilistic name and address cleaning and standardization. In: Australasian Data Mining Workshop. Canberra (2002)
69. Christen, P., Gayler, R.: Towards scalable real-time entity resolution using a similarity-aware inverted index approach. In: AusDM, CRPIT, vol. 87, pp. 51–60. Glenelg, Australia (2008)
70. Christen, P., Gayler, R., Hawking, D.: Similarity-aware indexing for real-time entity resolution. In: ACM CIKM, pp. 1565–1568. Hong Kong (2009)
71. Christen, P., Goiser, K.: Quality and complexity measures for data linkage and deduplication. In: F. Guillet, H. Hamilton (eds.) Quality Measures in Data Mining, *Studies in Computational Intelligence*, vol. 43, pp. 127–151. Springer (2007)
72. Christen, P., Pudjijono, A.: Accurate synthetic generation of realistic personal information. In: PAKDD, Springer LNAI, vol. 5476, pp. 507–514. Bangkok, Thailand (2009)
73. Churches, T.: A proposed architecture and method of operation for improving the protection of privacy and confidentiality in disease registers. BioMed Central Medical Research Methodology 3(1) (2003)
74. Churches, T., Christen, P.: Blind data linkage using n-gram similarity comparisons. In: PAKDD, Springer LNAI, vol. 3056, pp. 121–126. Sydney (2004)
75. Churches, T., Christen, P.: Some methods for blindfolded record linkage. BioMed Central Medical Informatics and Decision Making 4(9) (2004)
76. Churches, T., Christen, P., Lim, K., Zhu, J.X.: Preparation of name and address data for record linkage using hidden Markov models. BioMed Central Medical Informatics and Decision Making 2(9) (2002)
77. Cilibrasi, R., Vitányi, P.M.: Clustering by compression. IEEE Transactions on Information Theory 51(4), 1523–1545 (2005)
78. Clark, D.E.: Practical introduction to record linkage for injury research. Injury Prevention 10, 186–191 (2004)
79. Clifton, C., Kantarcioglu, M., Doan, A., Schadow, G., Vaidya, J., Elmagarmid, A., Suciu, D.: Privacy-preserving data integration and sharing. In: ACM SIGMOD workshop on Research issues in Data Mining and Knowledge Discovery, pp. 19–26 (2004)
80. Cochinwala, M., Kurien, V., Lalk, G., Shasha, D.: Efficient data reconciliation. Information Sciences 137(1–4), 1–15 (2001)
81. Cohen, W.: The WHIRL approach to data integration. IEEE Intelligent Systems 13(3), 20–24 (1998)
82. Cohen, W.: Data integration using similarity joins and a word-based information representation language. ACM Transactions on Information Systems 18(3), 288–321 (2000)
83. Cohen, W.: Integration of heterogeneous databases without common domains using queries based on textual similarity. In: ACM SIGMOD, pp. 201–212. Seattle (1998)
84. Cohen, W., Ravikumar, P., Fienberg, S.: A comparison of string distance metrics for name-matching tasks. In: Workshop on Information Integration on the Web, held at IJCAI, pp. 73–78. Acapulco (2003)
85. Cohen, W., Richman, J.: Learning to match and cluster large high-dimensional data sets for data integration. In: ACM SIGKDD, pp. 475–480. Edmonton (2002)

86. Conn, L., Bishop, G.: Exploring methods for creating a longitudinal census dataset. Tech. Rep. 1352.0.55.076, Australian Bureau of Statistics, Canberra (2005)
87. Curtis, A.J., Mills, J.W., Leitner, M.: Spatial confidentiality and GIS: Re-engineering mortality locations from published maps about Hurricane Katrina. International Journal of Health Geographics 5(1), 44–56 (2006)
88. Dal Bianco, G., Galante, R., Heuser, C.: A fast approach for parallel deduplication on multicore processors. In: ACM Symposium on Applied, Computing, pp. 1027–1032 (2011)
89. Damerau, F.J.: A technique for computer detection and correction of spelling errors. Communications of the ACM 7(3), 171–176 (1964)
90. Day, C.: Record linkage i: evaluation of commercially available record linkage software for use in NASS. Tech. Rep. STB Research Report STB-95-02, National Agricultural Statistics Service, Washington DC (1995)
91. Dey, D., Mookerjee, V., Liu, D.: Efficient techniques for online record linkage. IEEE Transactions on Knowledge and Data Engineering 23(3), 373–387 (2010)
92. Domingo-Ferrer, J., Torra, V.: Disclosure risk assessment in statistical microdata protection via advanced record linkage. Statistics and Computing 13(4), 343–354 (2003)
93. Dong, X., Halevy, A., Madhavan, J.: Reference reconciliation in complex information spaces. In: ACM SIGMOD, pp. 85–96. Baltimore (2005)
94. Draisbach, U., Naumann, F.: A comparison and generalization of blocking and windowing algorithms for duplicate detection. In: Workshop on Quality in Databases, held at VLDB. Lyon (2009)
95. Draisbach, U., Naumann, F.: Dude: The duplicate detection toolkit. In: Workshop on Quality in Databases, held at VLDB. Singapore (2010)
96. Du, W., Atallah, M., Kerschbaum, F.: Protocols for secure remote database access with approximate matching. In: First ACM Workshop on Security and Privacy in E-Commerce (2000)
97. Dunn, H.: Record linkage. American Journal of Public Health 36(12), 1412 (1946)
98. Durham, E., Xue, Y., Kantarcioglu, M., Malin, B.: Private medical record linkage with approximate matching. In: AMIA Annual Symposium Proceedings, p. 182. American Medical Informatics Association (2010)
99. Durham, E., Xue, Y., Kantarcioglu, M., Malin, B.: Quantifying the correctness, computational complexity, and security of privacy-preserving string comparators for record linkage. Information Fusion **In Press** (2011)
100. Durham, E.: A framework for accurate, efficient private record linkage. Ph.D. thesis, Faculty of the Graduate School of Vanderbilt University, Nashville, TN (2012)
101. Dwork, C.: Differential privacy. Automata, languages and programming pp. 1–12 (2006)
102. Elfeky, M.G., Verykios, V., Elmagarmid, A.K.: TAILOR: A record linkage toolbox. In: IEEE ICDE, pp. 17–28. San Jose (2002)
103. Elmagarmid, A.K., Ipeirotis, P.G., Verykios, V.: Duplicate record detection: A survey. IEEE Transactions on Knowledge and Data Engineering 19(1), 1–16 (2007)
104. Fagin, R., Naor, M., Winkler, P.: Comparing information without leaking it. Communications of the ACM 39(5), 77–85 (1996)
105. Faloutsos, C., Lin, K.I.: Fastmap: A fast algorithm for indexing, data-mining and visualization of traditional and multimedia datasets. In: ACM SIGMOD, pp. 163–174. San Jose (1995)
106. Fawcett T: ROC Graphs: Notes and practical considerations for researchers. Tech. Rep. HPL-2003-4, HP Laboratories, Palo Alto (2004)
107. Fellegi, I.P., Holt, D.: A systematic approach to automatic edit and imputation. Journal of the American Statistical Association pp. 17–35 (1976)
108. Fellegi, I.P., Sunter, A.B.: A theory for record linkage. Journal of the American Statistical Association 64(328), 1183–1210 (1969)
109. Fienberg, S.: Homeland insecurity: Datamining, terrorism detection, and confidentiality. Bull. Internat. Stat. Inst (2005)

110. Fienberg, S.: Privacy and confidentiality in an e-commerce world: Data mining, data warehousing, matching and disclosure limitation. Statistical Science **21**(2), 143–154 (2006)
111. Fogel, R.: New sources and new techniques for the study of secular trends in nutritional status, health, mortality, and the process of aging. NBER Historical Working Papers (1993)
112. Fortini, M., Liseo, B., Nuccitelli, A., Scanu, M.: On Bayesian record linkage. Research in Official Statistics **4**(1), 185–198 (2001)
113. Friedman, C., Sideli, R.: Tolerating spelling errors during patient validation. Computers and Biomedical Research **25**, 486–509 (1992)
114. Fu, Z., Christen, P., Boot, M.: Automatic cleaning and linking of historical census data using household information. In: Workshop on Domain Driven Data Mining, held at IEEE ICDM. Vancouver (2011)
115. Fu, Z., Christen, P., Boot, M.: A supervised learning and group linking method for historical census household linkage. In: AusDM, CRPIT, vol. 125. Ballarat, Australia (2011)
116. Fu, Z., Zhou, J., Christen, P., Boot, M.: Multiple instance learning for group record linkage. In: PAKDD, Springer LNAI. Kuala Lumpur, Malaysia (2012)
117. Galhardas, H., Florescu, D., Shasha, D., Simon, E.: An extensible framework for data cleaning. In: IEEE ICDE. San Diego (2000)
118. Gill, L.: OX-LINK: The Oxford medical record linkage system. In: Proc. IntGI Record Linkage Workshop and Exposition, pp. 15–33. Arlington, Virginia (1997)
119. Gill, L.: Methods for automatic record matching and linking and their use in national statistics. Tech. Rep. Methodology Series, no. 25, National Statistics, London (2001)
120. Giunchiglia, F., Yatskevich, M., Shvaiko, P.: Semantic matching: Algorithms and implementation. Journal on Data Semantics IX pp. 1–38 (2007)
121. Glasson, E., De Klerk, N., Bass, A., Rosman, D., Palmer, L., Holman, C.: Cohort profile: the Western Australian family connections genealogical project. International Journal of epidemiology **37**(1), 30–35 (2008)
122. Gliklich, R., Dreyer, N. (eds.): Registries for Evaluating Patient Outcomes: A UserGs Guide. No.10-EHC049. AHRQ, Publication (2010)
123. Goldreich, O.: Secure multi-party computation. Tech. rep., Department of Computer Science and Applied Mathematics, Weizmann Institute of Science, Israel (2002)
124. Gomatam, S., Carter, R., Ariet, M., Mitchell, G.: An empirical comparison of record linkage procedures. Statistics in Medicine **21**(10), 1485–1496 (2002)
125. Gong, R., Chan, T.K.: Syllable alignment: A novel model for phonetic string search. IEICE Transactions on Information and Systems **E89-D**(1), 332–339 (2006)
126. Grama, A., Karypis, G., Kumar, V., Gupta, A.: Introduction to parallel computing, 2 edn. Addison-Wesley Longman Publishing Co., Inc. (2003)
127. Gravano, L., Ipeirotis, P.G., Jagadish, H.V., Koudas, N., Muthukrishnan, S., Srivastava, D.: Approximate string joins in a database (almost) for free. In: VLDB, pp. 491–500. Roma (2001)
128. Gu, L., Baxter, R.: Adaptive filtering for efficient record linkage. In: SIAM international conference on data mining. Orlando, Florida (2004)
129. Gu, L., Baxter, R.: Decision models for record linkage. In: Selected Papers from AusDM, Springer LNCS 3755, pp. 146–160 (2006)
130. Guo, H., Zhu, H., Guo, Z., Zhang, X., Su, Z.: Address standardization with latent semantic association. In: ACM SIGKDD, pp. 1155–1164. Paris (2009)
131. Hajishirzi, H., Yih, W., Kolcz, A.: Adaptive near-duplicate detection via similarity learning. In: ACM SIGIR, pp. 419–426. Geneva, Switzerland (2010)
132. Hall, M., Frank, E., Holmes, G., Pfahringer, B., Reutemann, P., Witten, I.: The WEKA data mining software: an update. ACM SIGKDD Explorations **11**(1), 10–18 (2009)
133. Hall, P.A., Dowling, G.R.: Approximate string matching. ACM Computing Surveys **12**(4), 381–402 (1980)
134. Hall, R., Fienberg, S.: Privacy-preserving record linkage. In: Privacy in Statistical Databases, Springer LNCS 6344, pp. 269–283. Corfu, Greece (2010)

135. Han, J., Kamber, M.: Data mining: concepts and techniques, 2 edn. Morgan Kaufmann (2006)
136. Hand, D.: Classifier technology and the illusion of progress. Statistical Science **21**(1), 1–14 (2006)
137. Hassanzadeh, O., Miller, R.: Creating probabilistic databases from duplicated data. The VLDB Journal **18**(5), 1141–1166 (2009)
138. Heckerman, D.: Bayesian networks for data mining. Data mining and knowledge discovery **1**(1), 79–119 (1997)
139. Henzinger, M.: Finding near-duplicate web pages: a large-scale evaluation of algorithms. In: ACM SIGIR, pp. 284–291. Seattle (2006)
140. Hernandez, M.A., Stolfo, S.J.: The merge/purge problem for large databases. In: ACM SIGMOD, pp. 127–138. San Jose (1995)
141. Hernandez, M.A., Stolfo, S.J.: Real-world data is dirty: Data cleansing and the merge/purge problem. Data Mining and Knowledge Discovery **2**(1), 9–37 (1998)
142. Herschel, M., Naumann, F., Szott, S., Taubert, M.: Scalable iterative graph duplicate detection. IEEE Transactions on Knowledge and Data Engineering **X**(Y) (2011)
143. Herzog, T., Scheuren, F., Winkler, W.: Data quality and record linkage techniques. Springer Verlag (2007)
144. Hirsch, J.: An index to quantify an individual's scientific research output. Proceedings of the National Academy of Sciences of the United States of America **102**(46), 16,569–16,572 (2005)
145. Holmes, D., McCabe, C.M.: Improving precision and recall for Soundex retrieval. In: Proceedings of the IEEE International Conference on Information Technology—Coding and Computing. Las Vegas (2002)
146. Inan, A., Kantarcioglu, M., Bertino, E., Scannapieco, M.: A hybrid approach to private record linkage. In: IEEE ICDE, pp. 496–505 (2008)
147. Inan, A., Kantarcioglu, M., Ghinita, G., Bertino, E.: Private record matching using differential privacy. In: International Conference on Extending Database Technology, pp. 123–134 (2010)
148. Ioannou, E., Nejdl, W., Niederée, C., Velegrakis, Y.: On-the-fly entity-aware query processing in the presence of linkage. Proceedings of the VLDB Endowment **3**(1) (2010)
149. Jaro, M.A.: Advances in record-linkage methodology a applied to matching the 1985 Census of Tampa, Florida. Journal of the American Statistical Association **84**, 414–420 (1989)
150. Jentzsch, A., Isele, R., Bizer, C.: Silk-generating RDF links while publishing or consuming linked data. In: Poster at the International Semantic Web Conference. Shanghai (2010)
151. Jin, L., Li, C., Mehrotra, S.: Efficient record linkage in large data sets. In: DASFAA, pp. 137–146. Tokyo (2003)
152. Jokinen, P., Tarhio, J., Ukkonen, E.: A comparison of approximate string matching algorithms. Software—Practice and Experience **26**(12), 1439–1458 (1996)
153. Jonas, J., Harper, J.: Effective counterterrorism and the limited role of predictive data mining. Policy Analysis (584) (2006)
154. Jurczyk, P., Lu, J., Xiong, L., Cragan, J., Correa, A.: FRIL: A tool for comparative record linkage. In: AMIA Annual Symposium Proceedings, p. 440. American Medical Informatics Association (2008)
155. Kalashnikov, D., Mehrotra, S.: Domain-independent data cleaning via analysis of entity-relationship graph. ACM Transactions on Database Systems **31**(2), 716–767 (2006)
156. Kang, H., Getoor, L., Shneiderman, B., Bilgic, M., Licamele, L.: Interactive entity resolution in relational data: A visual analytic tool and its evaluation. IEEE Transactions on Visualization and Computer Graphics **14**(5), 999–1014 (2008)
157. Karakasidis, A., Verykios, V.: Privacy preserving record linkage using phonetic codes. In: Fourth Balkan Conference in Informatics, pp. 101–106. IEEE (2009)

158. Karakasidis, A., Verykios, V.: Advances in privacy preserving record linkage. In: E-activity and Innovative Technology, Advances in Applied Intelligence Technologies Book Series, pp. 22–34. IGI Global (2010)
159. Karakasidis, A., Verykios, V., Christen, P.: Fake injection strategies for private phonetic matching. In: International Workshop on Data Privacy Management. Leuven, Belgium (2011)
160. Kawai, H., Garcia-Molina, H., Benjelloun, O., Menestrina, D., Whang, E., Gong, H.: P-Swoosh: Parallel algorithm for generic entity resolution. Tech. Rep. 2006-19, Department of Computer Science, Stanford University (2006)
161. Kelman, C.W., Bass, J., Holman, D.: Research use of linked health data—A best practice protocol. Aust NZ Journal of Public Health **26**, 251–255 (2002)
162. Keskustalo, H., Pirkola, A., Visala, K., Leppanen, E., Jarvelin, K.: Non-adjacent digrams improve matching of cross-lingual spelling variants. In: String Processing and Information Retrieval, LNCS 2857, pp. 252–265. Manaus, Brazil (2003)
163. Kim, H., Lee, D.: Parallel linkage. In: ACM CIKM, pp. 283–292. Lisboa, Portugal (2007)
164. Kim, H., Lee, D.: Harra: fast iterative hashed record linkage for large-scale data collections. In: International Conference on Extending Database Technology, pp. 525–536. Lausanne, Switzerland (2010)
165. Kirsten, T., Kolb, L., Hartung, M., Gross, A., Köpcke, H., Rahm, E.: Data partitioning for parallel entity matching. Proceedings of the VLDB Endowment **3**(2) (2010)
166. Klenk, S., Thom, D., Heidemann, G.: The normalized compression distance as a distance measure in entity identification. Advances in Data Mining. Applications and Theoretical Aspects pp. 325–337 (2009)
167. Kolb, L., Thor, A., Rahm, E.: Multi-pass sorted neighborhood blocking with Map-Reduce. Computer Science-Research and, Development pp. 1–19 (2011)
168. Köpcke, H., Rahm, E.: Frameworks for entity matching: A comparison. Data and Knowledge Engineering **69**(2), 197–210 (2010)
169. Köpcke, H., Thor, A., Rahm, E.: Evaluation of entity resolution approaches on real-world match problems. Proceedings of the VLDB Endowment **3**(1–2), 484–493 (2010)
170. Koudas, N., Marathe, A., Srivastava, D.: Flexible string matching against large databases in practice. In: VLDB, pp. 1086–1094. Toronto (2004)
171. Krouse, W., Elias, B.: Terrorist Watchlist Checks and Air Passenger Prescreening. RL33645. Congressional Research Service (2009). CRS Report for Congress
172. Kukich, K.: Techniques for automatically correcting words in text. ACM Computing Surveys **24**(4), 377–439 (1992)
173. Kuzu, M., Kantarcioglu, M., Durham, E., Malin, B.: A constraint satisfaction cryptanalysis of Bloom filters in private record linkage. In: Privacy Enhancing Technologies, pp. 226–245. Springer (2011)
174. Lahiri, P., Larsen, M.: Regression analysis with linked data. Journal of the American statistical association **100**(469), 222–230 (2005)
175. Lait, A., Randell, B.: An assessment of name matching algorithms. Tech. rep., Department of Computer Science, University of Newcastle upon Tyne (1993)
176. Lee, D., Kang, J., Mitra, P., Giles, C.L., On, B.W.: Are your citations clean? Commununications of the ACM **50**, 33–38 (2007)
177. Lee, Y., Pipino, L., Funk, J., Wang, R.: Journey to data quality. The MIT Press (2009)
178. Lenzerini, M.: Data integration: A theoretical perspective. In: ACM SIGMOD-SIGACT-SIGART symposium on Principles of database systems, pp. 233–246. Madison (2002)
179. Li, P., Dong, X., Maurino, A., Srivastava, D.: Linking temporal records. Proceedings of the VLDB Endowment **4**(11) (2011)
180. Malin, B.: K-unlinkability: A privacy protection model for distributed data. Data and Knowledge Engineering **64**(1), 294–311 (2008)
181. Malin, B., Airoldi, E., Carley, K.: A network analysis model for disambiguation of names in lists. Computational and Mathematical Organization Theory **11**(2), 119–139 (2005)

182. Malin, B., Karp, D., Scheuermann, R.: Technical and policy approaches to balancing patient privacy and data sharing in clinical and translational research. Journal of investigative medicine: the official publication of the American Federation for Clinical Research **58**(1), 11 (2010)

183. Manghi, P., Mikulicic, M.: PACE: A general-purpose tool for authority control. Metadata and Semantic Research pp. 80–92 (2011)

184. McCallum, A., Nigam, K., Rennie, J., Seymore, K.: Automating the construction of Internet portals with machine learning. Information Retrieval **3**(2), 127–163 (2000)

185. McCallum, A., Nigam, K., Ungar, L.H.: Efficient clustering of high-dimensional data sets with application to reference matching. In: ACM SIGKDD, pp. 169–178. Boston (2000)

186. Menestrina, D., Benjelloun, O., Garcia-Molina, H.: Generic entity resolution with data confidences. In: First International VLDB Workshop on Clean Databases. Seoul, South Korea (2006)

187. Menestrina, D., Whang, S., Garcia-Molina, H.: Evaluating entity resolution results. Proceedings of the VLDB Endowment **3**(1–2), 208–219 (2010)

188. Michelson, M., Knoblock, C.A.: Learning blocking schemes for record linkage. In: AAAI. Boston (2006)

189. Mitchell, T.M.: Machine Learning. McGraw Hill (1997)

190. Monge, A.E.: Matching algorithms within a duplicate detection system. IEEE Data Engineering Bulletin **23**(4), 14–20 (2000)

191. Monge, A.E., Elkan, C.P.: The field-matching problem: Algorithm and applications. In: ACM SIGKDD, pp. 267–270. Portland (1996)

192. Moreau, E., Yvon, F., Cappé, O.: Robust similarity measures for named entities matching. In: 22nd International Conference on Computational Linguistics-Volume 1, pp. 593–600. Association for Computational Linguistics (2008)

193. Moustakides, G.V., Verykios, V.: Optimal stopping: A record linkage approach. Journal Data and Information Quality **1**, 9:1–9:34 (2009)

194. Narayanan, A., Shmatikov, V.: Myths and fallacies of personally identifiable information. Communications of the ACM **53**(6), 24–26 (2010)

195. Naumann, F., Herschel, M.: An introduction to duplicate detection, *Synthesis Lectures on Data Management*, vol. 3. Morgan and Claypool Publishers (2010)

196. Navarro, G.: A guided tour to approximate string matching. ACM Computing Surveys **33**(1), 31–88 (2001)

197. Newcombe, H., Kennedy, J.: Record linkage: making maximum use of the discriminating power of identifying information. Communications of the ACM **5**(11), 563–566 (1962)

198. Newcombe, H., Kennedy, J., Axford, S., James, A.: Automatic linkage of vital records. Science **130**(3381), 954–959 (1959)

199. Newcombe, H.B.: Handbook of record linkage: methods for health and statistical studies, administration, and business. Oxford University Press, Inc., New York, NY, USA (1988)

200. Nin, J., Muntes-Mulero, V., Martinez-Bazan, N., Larriba-Pey, J.L.: On the use of semantic blocking techniques for data cleansing and integration. In: IDEAS, pp. 190–198. Banff, Canada (2007)

201. Odell, M., Russell, R.: The soundex coding system. US Patents **1261167** (1918)

202. O'Keefe, C., Yung, M., Gu, L., Baxter, R.: Privacy-preserving data linkage protocols. In: ACM Workshop on Privacy in the Electronic Society, pp. 94–102. Washington DC (2004)

203. Okner, B.: Data matching and merging: An overview. NBER Chapters pp. 49–54 (1974)

204. On, B.W., Elmacioglu, E., Lee, D., Kang, J., Pei, J.: Improving grouped-entity resolution using quasi-cliques. In: IEEE ICDM, pp. 1008–1015 (2006)

205. On, B.W., Koudas, N., Lee, D., Srivastava, D.: Group linkage. In: IEEE ICDE, pp. 496–505. Istanbul (2007)

206. Oscherwitz, T.: Synthetic identity fraud: unseen identity challenge. Bank Security News **3**(7) (2005)

207. Pang, C., Gu, L., Hansen, D., Maeder, A.: Privacy-preserving fuzzy matching using a public reference table. Intelligent Patient Management pp. 71–89 (2009)
208. Patman, F., Thompson, P.: Names: A new frontier in text mining. In: ISI-2003, Springer LNCS 2665, pp. 27–38 (2003)
209. Paull, D.: A geocoded national address file for Australia: The G-NAF what, why, who and when? PSMA Australia Limited, Griffith, ACT, Australia (2003)
210. Pfeifer, U., Poersch, T., Fuhr, N.: Retrieval effectiveness of proper name search methods. Information Processing and Management 32(6), 667–679 (1996)
211. Philips, L.: The double-metaphone search algorithm. C/C++ User's Journal 18(6) (2000)
212. Phua, C., Smith-Miles, K., Lee, V., Gayler, R.: Resilient identity crime detection. IEEE Transactions on Knowledge and Data Engineering 24(3) (2012)
213. Poindexter, J., Popp, R., Sharkey, B.: Total information awareness (TIA). In: IEEE Aerospace Conference, 2003, vol. 6, pp. 2937–2944 (2003)
214. Pollock, J.J., Zamora, A.: Automatic spelling correction in scientific and scholarly text. Communications of the ACM 27(4), 358–368 (1984)
215. Porter, E.H., Winkler, W.E.: Approximate string comparison and its effect on an advanced record linkage system. Tech. Rep. RR97/02, US Bureau of the Census (1997)
216. Prabhakar, S., Shah, R., Singh, S.: Indexing uncertain data. In: C.C. Aggarwal (ed.) Managing and Mining Uncertain Data, Advances in Database Systems, vol. 35, pp. 299–325. Springer (2009)
217. Prasad, K., Faruquie, T., Joshi, S., Chaturvedi, S., Subramaniam, L., Mohania, M.: Data cleansing techniques for large enterprise datasets. In: SRII Global Conference, pp. 135–144. San Jose, USA (2009)
218. Pyle, D.: Data preparation for data mining. Morgan Kaufmann (1999)
219. Quantin, C., Bouzelat, H., Allaert, F., Benhamiche, A., Faivre, J., Dusserre, L.: How to ensure data quality of an epidemiological follow-up: Quality assessment of an anonymous record linkage procedure. International Journal of Medical Informatics 49(1), 117–122 (1998)
220. Quantin, C., Bouzelat, H., Allaert, F.A., Benhamiche, A.M., Faivre, J., Dusserre, L.: Automatic record hash coding and linkage for epidemiological follow-up data confidentiality. Methods of Information in Medicine 37(3), 271–277 (1998)
221. Quantin, C., Bouzelat, H., Dusserre, L.: Irreversible encryption method by generation of polynomials. Medical Informatics and the Internet in Medicine 21(2), 113–121 (1996)
222. Quass, D., Starkey, P.: Record linkage for genealogical databases. In: ACM SIGKDD Workshop on Data Cleaning, Record Linkage and Object Consolidation, pp. 40–42. Washington DC (2003)
223. Rabiner, L.: A tutorial on hidden Markov models and selected applications in speech recognition. Proceedings of the IEEE 77(2), 257–286 (1989)
224. Rahm, E., Do, H.H.: Data cleaning: Problems and current approaches. IEEE Data Engineering Bulletin 23(4), 3–13 (2000)
225. Rastogi, V., Dalvi, N., Garofalakis, M.: Large-scale collective entity matching. VLDB Endowment 4, 208–218 (2011)
226. Ravikumar, P., Cohen, W., Fienberg, S.: A secure protocol for computing string distance metrics. In: Workshop on Privacy and Security Aspects of Data Mining held at IEEE ICDM, pp. 40–46. Brighton, UK (2004)
227. Ruggles, S.: Linking historical censuses: A new approach. History and Computing 14(1–2), 213–224 (2002)
228. Rushton, G., Armstrong, M., Gittler, J., Greene, B., Pavlik, C., West, M., Zimmerman, D.: Geocoding in cancer research: A review. American Journal of Preventive Medicine 30(2), S16–S24 (2006)
229. Sadinle, M., Hall, R., Fienberg, S.: Approaches to multiple record linkage. Proceedings of International Statistical Institute (2011)

230. Sarawagi, S.: Information extraction. Foundations and Trends in Databases **1**(3), 261–377 (2008)
231. Sarawagi, S., Bhamidipaty, A.: Interactive deduplication using active learning. In: ACM SIGKDD, pp. 269–278. Edmonton (2002)
232. Sarawagi, S., Kirpal, A.: Efficient set joins on similarity predicates. In: ACM SIGMOD, pp. 754–765. Paris (2004)
233. Sariyar, M., Borg, A.: The RecordLinkage package: Detecting errors in data. The R Journal **2**(2), 61–67 (2010)
234. Scannapieco, M., Figotin, I., Bertino, E., Elmagarmid, A.: Privacy preserving schema and data matching. In: ACM SIGMOD, pp. 653–664 (2007)
235. Scheuren, F., Winkler, W.: Regression analysis of data files that are computer matched. Statistics of income: Turning administrative systems into information systems **1299**(1), 131 (1993)
236. Schewe, K., Wang, Q.: On the decidability and complexity of identity knowledge representation. In: Database Systems for Advanced Applications, Springer LNCS 7238, pp. 288–302. Busan, South Korea (2012)
237. Schneier, B.: Applied cryptography: Protocols, algorithms, and source code in C, 2 edn. John Wiley and Sons, Inc., New York (1996)
238. Schnell, R., Bachteler, T., Bender, S.: A toolbox for record linkage. Austrian Journal of Statistics **33**(1& 2), 125–133 (2004)
239. Schnell, R., Bachteler, T., Reiher, J.: Privacy-preserving record linkage using Bloom filters. BioMed Central Medical Informatics and Decision Making **9**(1) (2009)
240. Seymore, K., McCallum, A., Rosenfeld, R.: Learning hidden Markov model structure for information extraction. In: AAAI Workshop on Machine Learning for Information Extraction, pp. 37–42 (1999)
241. Smith, M., Newcombe, H.: Methods for computer linkage of hospital admission-separation records into cumulative health histories. Methods of Information in Medicine **14**(3), 118–125 (1975)
242. Smith, M., Newcombe, H.: Accuracies of computer versus manual linkages of routine health records. Methods of Information in Medicine **18**(2), 89–97 (1979)
243. Snae, C.: A comparison and analysis of name matching algorithms. International Journal of Applied Science, Engineering and Technology **4**(1), 252–257 (2007)
244. Song, D., Wagner, D., Perrig, A.: Practical techniques for searches on encrypted data. In: IEEE Symposium on Security and Privacy, pp. 44–55 (2000)
245. Su, W., Wang, J., Lochovsky, F.H.: Record matching over query results from multiple web databases. IEEE Transactions on Knowledge and Data Engineering **22**(4), 578–589 (2009)
246. Summerhayes, R., Holder, P., Beard, J., Morgan, G., Christen, P., Willmore, A., Churches, T.: Automated geocoding of routinely collected health data in New South Wales. New South Wales Public Health Bulletin **17**(4), 33–38 (2006)
247. Sweeney, L.: Computational disclosure control: A primer on data privacy protection. Ph.D. thesis, Massachusetts Institute of Technology, Dept. of Electrical Engineering and Computer Science (2001)
248. Sweeney, L.: K-anonymity: A model for protecting privacy. International Journal of Uncertainty Fuzziness and Knowledge Based Systems **10**(5), 557–570 (2002)
249. Talburt, J.: Entity Resolution and Information Quality. Morgan Kaufmann (2011)
250. Talburt, J.R., Zhou, Y., Shivaiah, S.Y.: SOG: A synthetic occupancy generator to support entity resolution instruction and research. In: International Conference on Information Quality, pp. 91–105. Potsdam, Germany (2009)
251. Technologies, M.: AutoStan and AutoMatch, User's Manuals (1998). Kennebunk, Maine
252. Tejada, S., Knoblock, C.A., Minton, S.: Learning domain-independent string transformation weights for high accuracy object identification. In: ACM SIGKDD, pp. 350–359. Edmonton (2002)

253. Torra, V., Domingo-Ferrer, J.: Record linkage methods for multidatabase data mining. Studies in Fuzziness and Soft Computing **123**, 101–132 (2003)
254. Torra, V., Domingo-Ferrer, J., Torres, A.: Data mining methods for linking data coming from several sources. In: Third Joint UN/ECE-Eurostat Work Session on Statistical Data Confidentiality, Eurostat. Monographs in Official Statistics. Luxembourg (2004)
255. Trepetin, S.: Privacy-preserving string comparisons in record linkage systems: a review. Information Security Journal: A Global Perspective **17**(5), 253–266 (2008)
256. US Federal Geographic Data Committee. Homeland Security and Geographic Information Systems: How GIS and mapping technology can save lives and protect property in post-September 11th America. Public Health GIS News and, Information (52), 21–23 (2003)
257. Vaidya, J., Clifton, C., Zhu, M.: Privacy preserving data mining, vol. 19. Springer (2006)
258. Van Berkel, B., De Smedt, K.: Triphone analysis: A combined method for the correction of orthographical and typographical errors. In: Second Conference on Applied Natural Language Processing, pp. 77–83. Austin (1988)
259. Vapnik, V.: The nature of statistical learning theory. Springer (2000)
260. Vatsalan, D., Christen, P., Verykios, V.: An efficient two-party protocol for approximate matching in private record linkage. In: AusDM, CRPIT, vol. 121. Ballarat, Australia (2011)
261. Verykios, V., Elmagarmid, A., Houstis, E.: Automating the approximate record-matching process. Information Sciences **126**(1–4), 83–98 (2000)
262. Verykios, V., Karakasidis, A., Mitrogiannis, V.: Privacy preserving record linkage approaches. Int. J. of Data Mining, Modelling and Management **1**(2), 206–221 (2009)
263. Verykios, V., George, M.V., Elfeky, M.G.: A Bayesian decision model for cost optimal record matching. The VLDB Journal **12**(1), 28–40 (2003)
264. Volz, J., Bizer, C., Gaedke, M., Kobilarov, G.: Silk—a link discovery framework for the web of data. In: Second Linked Data on the Web Workshop (2009)
265. de Vries, T., Ke, H., Chawla, S., Christen, P.: Robust record linkage blocking using suffix arrays. In: ACM CIKM, pp. 305–314. Hong Kong (2009)
266. de Vries, T., Ke, H., Chawla, S., Christen, P.: Robust record linkage blocking using suffix arrays and Bloom filters. ACM Transactions on Knowledge Discovery from Data **5**(2) (2011)
267. Wang, G., Chen, H., Atabakhsh, H.: Automatically detecting deceptive criminal identities. Communications of the ACM **47**(3), 70–76 (2004)
268. Wartell, J., McEwen, T.: Privacy in the information age: A guide for sharing crime maps and spatial data. Institute for Law and Justice, NCJ 188739 (2001)
269. Weis, M., Naumann, F.: Detecting duplicate objects in xml documents. In: International Workshop on Information Quality in Information Systems, pp. 10–19. Paris (2004)
270. Weis, M., Naumann, F.: Dogmatix tracks down duplicates in XML. In: ACM SIGMOD, pp. 431–442. Baltimore (2005)
271. Weis, M., Naumann, F., Brosy, F.: A duplicate detection benchmark for XML (and relational) data. In: Workshop on Information Quality for Information Systems (IQIS). Chicago (2006)
272. Weis, M., Naumann, F., Jehle, U., Lufter, J., Schuster, H.: Industry-scale duplicate detection. Proceedings of the VLDB Endowment **1**(2), 1253–1264 (2008)
273. West, D.: Introduction to graph theory, 3 edn. Prentice Hall (2007)
274. Whang, S., Garcia-Molina, H.: Entity resolution with evolving rules. Proceedings of the VLDB Endowment **3**(1–2), 1326–1337 (2010)
275. Whang, S.E., Garcia-Molina, H.: Developments in generic entity resolution. IEEE Data Engineering Bulletin **34**(3), 51–59 (2011)
276. Whang, S.E., Garcia-Molina, H.: Joint entity resolution. In: IEEE ICDE. Arlington, Virginia (2012)
277. Whang, S.E., Menestrina, D., Koutrika, G., Theobald, M., Garcia-Molina, H.: Entity resolution with iterative blocking. In: ACM SIGMOD, pp. 219–232. Providence, Rhode Island (2009)
278. Williams, G.J.: Rattle: a data mining GUI for R. The R Journal **1**(2), 45–55 (2009)

279. Winkler, W.: String comparator metrics and enhanced decision rules in the Fellegi-Sunter model of record linkage. In: Proceedings of the Section on Survey Research Methods, pp. 354–359. American Statistical Association (1990)

280. Winkler, W.E.: Using the EM algorithm for weight computation in the Fellegi-Sunter model of record linkage. Tech. Rep. RR2000/05, US Bureau of the Census, Washington, DC (2000)

281. Winkler, W.E.: Methods for record linkage and Bayesian networks. Tech. Rep. RR2002/05, US Bureau of the Census, Washington, DC (2001)

282. Winkler, W.E.: Record linkage software and methods for merging administrative lists. Tech. Rep. RR2001/03, US Bureau of the Census, Washington, DC (2001)

283. Winkler, W.E.: Approximate string comparator search strategies for very large administrative lists. Tech. Rep. RR2005/02, US Bureau of the Census, Washington, DC (2005)

284. Winkler, W.E.: Overview of record linkage and current research directions. Tech. Rep. RR2006/02, US Bureau of the Census, Washington, DC (2006)

285. Winkler, W.E.: Automatic estimation of record linkage false match rates. Tech. Rep. RR2007/05, US Bureau of the Census, Washington, DC (2007)

286. Winkler, W.E., Thibaudeau, Y.: An application of the Fellegi-Sunter model of record linkage to the 1990 U.S. decennial census. Tech. Rep. RR1991/09, US Bureau of the Census, Washington, DC (1991)

287. Winkler, W.E., Yancey, W.E., Porter, E.H.: Fast record linkage of very large files in support of decennial and administrative records projects. In: Proceedings of the Section on Survey Research Methods, pp. 2120–2130. American Statistical Association (2010)

288. Witten, I.H., Moffat, A., Bell, T.C.: Managing Gigabytes, 2 edn. Morgan Kaufmann (1999)

289. Xiao, C., Wang, W., Lin, X.: Ed-join: an efficient algorithm for similarity joins with edit distance constraints. Proceedings of the VLDB Endowment 1(1), 933–944 (2008)

290. Yakout, M., Atallah, M., Elmagarmid, A.: Efficient private record linkage. In: IEEE ICDE, pp. 1283–1286 (2009)

291. Yakout, M., Elmagarmid, A., Elmeleegy, H., Ouzzani, M., Qi, A.: Behavior based record linkage. Proceedings of the VLDB Endowment 3(1–2), 439–448 (2010)

292. Yan, S., Lee, D., Kan, M.Y., Giles, L.C.: Adaptive sorted neighborhood methods for efficient record linkage. In: ACM/IEEE-CS joint conference on Digital Libraries, pp. 185–194 (2007)

293. Yancey, W.E.: An adaptive string comparator for record linkage. Tech. Rep. RR2004/02, US Bureau of the Census (2004)

294. Yancey, W.E.: Evaluating string comparator performance for record linkage. Tech. Rep. RR2005/05, US Bureau of the Census (2005)

295. Yancey, W.E.: BigMatch: A program for extracting probable matches from a large file for record linkage. Tech. Rep. RRC2007/01, US Bureau of the Census (2007)

296. Yu, P., Han, J., Faloutsos, C.: Link Mining: Models, Algorithms, and Applications. Springer (2010)

297. Zaki, M., Ho, C.: Large-scale parallel data mining. Springer LNCS 1759 (2000)

298. Zhang, Y., Lin, X., Zhang, W., Wang, J., Lin, Q.: Effectively indexing the uncertain space. IEEE Transactions on Knowledge and Data Engineering 22(9), 1247–1261 (2010)

299. Zhao, H.: Semantic matching across heterogeneous data sources. Communications of the ACM 50(1), 45–50 (2007)

300. Zhu, J.J., Ungar, L.H.: String edit analysis for merging databases. In: KDD workshop on text mining, held at ACM SIGKDD. Boston (2000)

301. Zingmond, D., Ye, Z., Ettner, S., Liu, H.: Linking hospital discharge and death records—accuracy and sources of bias. Journal of Clinical Epidemiology 57, 21–29 (2004)

302. Zobel, J., Dart, P.: Phonetic string matching: Lessons from information retrieval. In: ACM SIGIR, pp. 166–172. Zürich, Switzerland (1996)

303. Zobel, J., Moffat, A.: Inverted files for text search engines. ACM Computing Surveys 38(2), 6 (2006)

Index

Index